Mesh Methods

Mesh Methods—Numerical Analysis and Experiments

Editors

Viktor A. Rukavishnikov
Pedro M. Lima
Ildar B. Badriev

MDPI • Basel • Beijing • Wuhan • Barcelona • Belgrade • Manchester • Tokyo • Cluj • Tianjin

Editors
Viktor A. Rukavishnikov
Russian Academy of Sciences
Russia

Pedro M. Lima
University of Lisbon
Portugal

Ildar B. Badriev
Kazan Federal University
Russia

Editorial Office
MDPI
St. Alban-Anlage 66
4052 Basel, Switzerland

This is a reprint of articles from the Special Issue published online in the open access journal *Symmetry* (ISSN 2073-8994) (available at: http://www.mdpi.com).

For citation purposes, cite each article independently as indicated on the article page online and as indicated below:

LastName, A.A.; LastName, B.B.; LastName, C.C. Article Title. *Journal Name* **Year**, *Volume Number*, Page Range.

ISBN 978-3-0365-0376-9 (Hbk)
ISBN 978-3-0365-0377-6 (PDF)

© 2021 by the authors. Articles in this book are Open Access and distributed under the Creative Commons Attribution (CC BY) license, which allows users to download, copy and build upon published articles, as long as the author and publisher are properly credited, which ensures maximum dissemination and a wider impact of our publications.

The book as a whole is distributed by MDPI under the terms and conditions of the Creative Commons license CC BY-NC-ND.

Contents

About the Editors vii

Preface to "Mesh Methods—Numerical Analysis and Experiments" ix

Samsul Ariffin Abdul Karim, Azizan Saaban and Van Thien Nguyen
Scattered Data Interpolation Using Quartic Triangular Patch for Shape-Preserving Interpolation and Comparison with Mesh-Free Methods
Reprinted from: *Symmetry* **2020**, *12*, 1071, doi:10.3390/sym12071071 1

Nikhil Anand, Neda Ebrahimi Pour, Harald Klimach and Sabine Roller
Utilization of the Brinkman Penalization to Represent Geometries in a High-Order Discontinuous Galerkin Scheme on Octree Meshes
Reprinted from: *Symmetry* **2019**, *11*, 1126, doi:10.3390/sym11091126 27

Victor A. Rukavishnikov and Elena I. Rukavishnikova
Numerical Method for Dirichlet Problem with Degeneration of the Solution on the Entire Boundary
Reprinted from: *Symmetry* **2019**, *11*, 1455, doi:10.3390/sym11121455 49

Yuhui Chen, Guoshuai Zhang, Ruolin Zhang, Timothy Gupta and Ahmed Katayama
Finite Element Study on the Wear Performance of Movable Jaw Plates of Jaw Crushers after a Symmetrical Laser Cladding Path
Reprinted from: *Symmetry* **2020**, *12*, 1126, doi:10.3390/sym11091126 61

Ratinan Boonklurb, Ampol Duangpan and Phansphitcha Gugaew
Numerical Solution of Direct and Inverse Problems for Time-Dependent Volterra Integro-Differential Equation Using Finite Integration Method with Shifted Chebyshev Polynomials
Reprinted from: *Symmetry* **2020**, *12*, 497, doi:10.3390/sym12040497 73

Galina Muratova, Tatiana Martynova, Evgeniya Andreeva, Vadim Bavin and Zeng-Qi Wang
Numerical Solution of the Navier–Stokes Equations Using Multigrid Methods with HSS-Based and STS-Based Smoothers
Reprinted from: *Symmetry* **2020**, *12*, 233, doi:10.3390/sym12020233 93

Khudija Bibi and Tooba Feroze
Discrete Symmetry Group Approach for Numerical Solution of the Heat Equation
Reprinted from: *Symmetry* **2020**, *12*, 359, doi:10.3390/sym12030359 105

About the Editors

Viktor A. Rukavishnikov

Pedro M. Lima

Ildar B. Badriev

Preface to "Mesh Methods—Numerical Analysis and Experiments"

Mathematical models of various natural processes are described by differential equations, systems of partial differential equations and integral equations. In most cases, the exact solution to such problems cannot be determined; therefore, one has to use grid methods to calculate an approximate solution using high-performance computing systems. These methods include the finite element method, the finite difference method, the finite volume method and combined methods. In this Special Issue, we bring to your attention works on theoretical studies of grid methods for approximation, stability and convergence, as well as the results of numerical experiments confirming the effectiveness of the developed methods. Of particular interest are new methods for solving boundary value problems with singularities, the complex geometry of the domain boundary and nonlinear equations. A part of the articles is devoted to the analysis of numerical methods developed for calculating mathematical models in various fields of applied science and engineering applications. As a rule, the ideas of symmetry are present in the design schemes and make the process harmonious and efficient.

Viktor A. Rukavishnikov, Pedro M. Lima, Ildar B. Badriev
Editors

Article

Scattered Data Interpolation Using Quartic Triangular Patch for Shape-Preserving Interpolation and Comparison with Mesh-Free Methods

Samsul Ariffin Abdul Karim [1,*], Azizan Saaban [2] and Van Thien Nguyen [3]

1. Fundamental and Applied Sciences Department and Centre for Smart Grid Energy Research (CSMER), Institute of Autonomous System, Universiti Teknologi PETRONAS, Bandar Seri Iskandar, Seri Iskandar 32610, Malaysia
2. School of Quantitative Sciences, UUMCAS, Universiti Utara Malaysia, Kedah 06010, Malaysia; azizan.s@uum.edu.my
3. FPT University, Education Zone, Hoa Lac High Tech Park, Km29 Thang Long Highway, Thach That Ward, Hanoi 10000, Vietnam; ThienNV15@fe.edu.vn
* Correspondence: samsul_ariffin@utp.edu.my

Received: 30 March 2020; Accepted: 1 May 2020; Published: 30 June 2020

Abstract: Scattered data interpolation is important in sciences, engineering, and medical-based problems. Quartic Bézier triangular patches with 15 control points (ordinates) can also be used for scattered data interpolation. However, this method has a weakness; that is, in order to achieve C^1 continuity, the three inner points can only be determined using an optimization method. Thus, we cannot obtain the exact Bézier ordinates, and the quartic scheme is global and not local. Therefore, the quartic Bézier triangular has received less attention. In this work, we use Zhu and Han's quartic spline with ten control points (ordinates). Since there are only ten control points (as for cubic Bézier triangular cases), all control points can be determined exactly, and the optimization problem can be avoided. This will improve the presentation of the surface, and the process to construct the scattered surface is local. We also apply the proposed scheme for the purpose of positivity-preserving scattered data interpolation. The sufficient conditions for the positivity of the quartic triangular patches are derived on seven ordinates. We obtain nonlinear equations that can be solved using the regula-falsi method. To produce the interpolated surface for scattered data, we employ four stages of an algorithm: (a) triangulate the scattered data using Delaunay triangulation; (b) assign the first derivative at the respective data; (c) form a triangular surface via convex combination from three local schemes with C^1 continuity along all adjacent triangles; and (d) construct the scattered data surface using the proposed quartic spline. Numerical results, including some comparisons with some existing mesh-free schemes, are presented in detail. Overall, the proposed quartic triangular spline scheme gives good results in terms of a higher coefficient of determination (R^2) and smaller maximum error (Max Error), requires about 12.5% of the CPU time of the quartic Bézier triangular, and is on par with Shepard triangular-based schemes. Therefore, the proposed scheme is significant for use in visualizing large and irregular scattered data sets. Finally, we tested the proposed positivity-preserving interpolation scheme to visualize coronavirus disease 2019 (COVID-19) cases in Malaysia.

Keywords: quartic spline; triangulation; scattered data; continuity; surface reconstruction; positivity-preserving; interpolation

1. Introduction

Scattered data interpolation and approximation are still active research topics in computer-aided design (CAD) and geometric modeling [1–9]. This is because engineers and scientists often face the problem of how to produce smooth curves and surfaces for the raw data obtained from experiments or observations. This is where scattered data interpolation can be used to assist them. To construct smooth curves and surfaces, some mathematical formulations are required. This can be achieved using functions which are well-established, such as the Bézier, B-spline, and radial basis functions (RBFs). All these methods are guaranteed to produce smooth curves and surfaces.

The formulation problem in scattered data interpolation can be described as follows:

Given functional data

$$(x_i, y_i, z_i), \quad i = 1, 2, \ldots, N$$

construct a smooth C^1 surface $z = F(x, y)$ such that

$$z_i = F(x_i, y_i), \quad i = 1, 2, \ldots, N$$

To solve the above problem, there are many methods that can be used, such as meshless methods (e.g., radial basis functions (RBFs) and many types of Shepard's families). However, some meshless schemes are global. Fasshauer [10] gave details on many meshless methods to solve the problems arising in scattered data interpolation and approximation, as well as partial differential equations. Beyond that, another approach that can be used to solve the problem is the triangulation of the given data points. Then, the Bézier or spline triangular can be used to construct a piecewise smooth surface with some degree of smoothness, such as C^1 or C^2. The Shepard triangular can also be used to produce a continuous surface from irregular scattered data. For instance, Cavoretto et al. [6], Dell'Accio and Di Tommaso [11], and Dell'Accio et al. [12,13] have discussed the application of the Shepard triangular for surface reconstruction. However, their schemes require more computation time in order to produce the interpolated surfaces.

Crivellaro et al. [14] applied RBFs to reconstruct 3D scattered data via new algorithms, which involves an adaptive multi-level interpolation approach based on implicit surface representation. The least squares approximation is used to remove the noise that appears in the scattered data. Chen and Cao [15] employed neural network operators of a logistic function through translations and dilation. Meanwhile, Bracco et al. [2] considered scattered data fitting using hierarchical splines where the local solutions are represented in variable degrees of the polynomial spline. Zhou and Li [16] studied scattered noise data by extending the weighted least squares method via triangulating the data points. Zhou and Li [17] discussed the scattered data interpolation for noisy data by using bivariate splines defined on triangulation. Qian et al. [18] also considered scattered data interpolation by using a new recursive algorithm based on the non-tensor product of bivariate splines. Liu [19] constructed local multilevel scattered data interpolation by proposing a new idea (i.e., nested scattered data sets), and scaled the compactly supported RBFs. Borne and Wende [3] also considered the meshless scheme based on definite RBFs for scattered data interpolation. In their study, they applied the domain decomposition methods to produce a symmetric-saddle point system. Joldes et al. [20] modified the moving least squares (MLS) methods by integrating the polynomial bases to solve the scattered data interpolation problem. Brodlie et al. [5] discussed the constrained surface interpolation by using the Shepard interpolant. The solution to the problem is obtained by solving some optimization. Lai and Meile [21] discussed scattered data interpolation by using nonnegative bivariate triangular splines to preserve the shape of the scattered data. Schumaker and Speleers [22] considered the nonnegativity preservation of scattered data by using macro-element spline spaces including Clough–Tocher macro-elements. Furthermore, they also give general results for range-restricted interpolation. Karim et al. [23] discussed the spatial interpolation for rainfall data by employing cubic Bézier triangular patches to interpolate the scattered data. Karim et al. [24] have constructed a new type of cubic Bézier-like triangular patches for scattered data interpolation. Karim and Saaban [25] constructed the terrain data using cubic Ball triangular patches [23]. In this study, they show that the scattered data interpolation scheme by Said and Rahmat [26] is not C^1 everywhere. Thus, a new condition for C^1 continuity is derived. The final surface is C^1 and provides a smooth surface. Feng and Zhang [27] proposed piecewise bivariate Hermite interpolations based on triangulation.

They applied the scheme for large scattered data sets to produce high-accuracy surface reconstruction. Sun et al. [28] constructed bivariate rational interpolation defined on a triangular domain for scattered data lying on a parallel line. They only considered a few data sets, and it was not tested for large data sets. By using a rational spline, the computation time increases. Bozzini et al. [4] proposed a polyharmonic spline to approximate the noisy scattered data.

The main motivation of the present study is described in the following paragraphs. In triangulation-based approaches to scattered data interpolation, cubic Bézier triangular or quintic Bézier triangular patches are the common methods. The quartic Bézier triangular has received less attention due to the need to solve optimization problems in order to calculate the Bézier ordinates. This approach increases the computation time. There are four steps in constructing a surface using a triangulation method: (a) triangulate the domain by using Delaunay triangulation; (b) specify the first partial derivative at the data points (sites); (c) assign the control points or ordinates for each triangular patch; and finally (d) the surface is constructed via a convex combination scheme. Goodman and Said [29] constructed a suitable C^1 triangular interpolant for scattered data interpolation using a convex combination scheme between three local schemes. Their work is different from that of Foley and Opitz [30]. However, both studies developed a C^1 cubic triangular convex combination scheme. Said and Rahmat [26] constructed a scattered data surface using cubic Ball triangular patches [31,32] with the same approach as in Goodman and Said [29]. Based on the numerical results, their scheme gave the same results as cubic Bézier triangular patches. The main advantages of cubic Ball triangular patches are that the required computation is 7% less when compared with the work of Goodman and Said [29]. This is what has been claimed by References [26,29]. However, in the work of Karim and Saaban [25], it was proved that Said and Rahmat [26] is not C^1 continuous everywhere, and Karim and Saaban [25] found that the [26] scheme produced the same surface for scattered data interpolation when the inner coefficient was calculated by using Reference [29]. Hussain and Hussain [33] proposed the rational cubic Bézier triangular for positivity-preserving scattered data interpolation. They claimed that their proposed scheme is C^1 positive everywhere. However, from their results, it is possible that their scheme may not be positive everywhere. Chan and Ong [7] considered range-restricted interpolation using a cubic Bézier triangular comprising three local schemes. All the schemes were implemented by estimating the partial derivatives at the respective knots using the method proposed by Goodman et al. [34].

Other than the use of cubic Ball and cubic Bézier triangular patches for scattered data interpolation, there are some studies that have utilized quartic Bézier triangular and rational quartic Bézier triangular patches for scattered data interpolation. For instance, Saaban et al. [35] constructed C^1 (or G^1) scattered data interpolation based on the quartic Bézier triangular. Piah et al. [36] considered C^1 range-restricted positivity-preserving scattered data interpolation by using the quartic Bézier triangular. They employed an optimization method (i.e., the minimized sum of squares) to calculate the inner Bézier points proposed in Saaban et al. [35]. Hussain et al. [37] extended this idea to construct convexity-preserving scattered data interpolation. Hussain et al. [38] constructed a new scattered data interpolation scheme by using the rational quartic Bézier triangular. They applied it to positivity-preserving interpolation. However, to achieve C^1 continuity, we still need to solve some optimization problems. This is the main weakness of quartic Bézier triangular patches when applied to scattered data interpolation. Some good surveys on scattered data interpolation can be found in [39–43].

The present study aims to answer the following research questions:

a. Can we construct a scattered data interpolation scheme by using quartic triangular patches but without an optimization method?

b. How can we produce a C^1 surface (everywhere)?

c. Is the proposed scheme better than some existing schemes in terms of CPU time, coefficient of determination (R^2), and maximum error?

To answer these research questions, we will use the quartic triangular basis initiated by Zhu and Han [44]. The main advantage of using this quartic basis is that it only requires ten control points to construct one triangular patch. This is the same as the number of control points in the cubic Bézier triangular patch. Thus, in order to construct C^1 scattered data interpolation using the quartic spline triangular, we can employ the Foley and Opitz [30] cubic precision scheme to calculate the inner ordinates. With this, the optimization problem required in a quartic triangular basis will be avoided. Hence, this will show that the proposed scheme is local. Furthermore, the proposed scheme is different from the works of Lai and Meile [21] and Schumaker and Spellers [22], even though all schemes required triangulation of the given data in the first step.

Some contributions from the present study are described below:

1. The proposed scattered data interpolation scheme produces a C^1 surface without any optimization method like Piah et al. [36], Saaban et al. [35] and Hussain et al. [37,38].

2. The proposed scheme is local; meanwhile, the schemes presented in Piah et al. [36], Saaban et al. [35] and Hussain et al. [37,38] are global.

3. Based on the CPU time needed to construct the surface, the proposed scheme is faster than quartic Bézier triangular patches. Thus, the reconstruction of scattered surfaces from large data sets can be performed in less time.

4. Furthermore, the proposed positivity-preserving scattered data interpolation is capable of producing a better interpolated surface than quartic Bézier triangular patches. This lies in contrast to scattered data schemes by Ali et al. [1], Draman et al. [9] and Karim et al. [24], which are not positivity-preserving interpolations.

This paper is organized as follows: In Section 2 we give a review of the triangular basis initiated by Zhu and Han [44], and the derivation of the quartic triangular basis with ten control points. Some graphical results are presented, as well as the construction of a local scheme comprising convex combination between three local schemes. The numerical results and the discussion are given in Section 3 with various numerical and graphical results, including a comparison with some existing schemes. Error analysis is also investigated in this section. The construction of the positive scattered data interpolant is discussed in Section 4. Meanwhile, numerical results for positivity-preserving scattered data interpolation are shown in Section 5. Conclusions and future recommendations are given in the final section.

2. Materials and Methods

2.1. Review of the Cubic Triangular Bases of Zhu And Han

Zhu and Han [44] proposed a new cubic triangular basis with three exponential parameters α, β, γ. Since we are dealing with triangulation, the barycentric coordinate (u, v, w) on the triangle T_1 with vertices V_1, V_2 and V_3 is defined by $u + v + w = 1$, where $u, v, w \geq 0$. Set the point inside the triangle as $V(x, y) \in R^2$ (as shown in Figure 1), which can be expressed as:

$$V = uV_1 + vV_2 + wV_3 \tag{1}$$

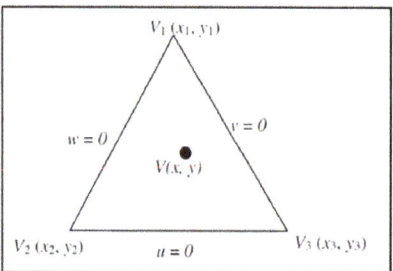

Figure 1. Triangle.

Definition 1. Let the parameters $\alpha, \beta, \gamma \in [2, \infty]$ and the triangular domain $D = \{(u, v, w) | u + v + w = 1\}$; the following are cubic Bernstein–Bézier basis functions ([44]):

$$\begin{aligned}
\{B^3_{3,0,0}(u,v,w;\alpha,\beta,\gamma) &= u^\alpha, & B^3_{0,3,0}(u,v,w;\alpha,\beta,\gamma) = v^\beta, & B^3_{0,0,3}(u,v,w;\alpha,\beta,\gamma) \\
&= w^\gamma, & B^3_{2,1,0}(u,v,w;\alpha,\beta,\gamma) & \\
&= u^2 v \left[\tfrac{3-2u-u^{\alpha-2}}{1-u}\right], & B^3_{2,0,1}(u,v,w;\alpha,\beta,\gamma) & \\
&= u^2 w \left[\tfrac{3-2u-u^{\alpha-2}}{1-u}\right], & B^3_{1,2,0}(u,v,w;\alpha,\beta,\gamma) & \\
&= uv^2 \left[\tfrac{3-2v-v^{\beta-2}}{1-v}\right], & B^3_{0,2,1}(u,v,w;\alpha,\beta,\gamma) & \\
&= v^2 w \left[\tfrac{3-2v-v^{\beta-2}}{1-v}\right], & B^3_{1,0,2}(u,v,w;\alpha,\beta,\gamma) & \\
&= uw^2 \left[\tfrac{3-2w-w^{\gamma-2}}{1-w}\right], & B^3_{0,1,2}(u,v,w;\alpha,\beta,\gamma) & \\
&= vw^2 \left[\tfrac{3-2w-w^{\gamma-2}}{1-w}\right], & B^3_{1,1,1}(u,v,w;\alpha,\beta,\gamma) = 6uvw.
\end{aligned} \qquad (2)$$

Zhu and Han's triangular basis functions satisfy the following properties:

Non-negativity:
$$B^3_{i,j,k}(u,v,w;\alpha,\beta,\gamma) \geq 0, \quad i+j+k=3.$$

Partition of unity:
$$\sum_{i+j+k=3} B^3_{i,j,k}(u,v,w;\alpha,\beta,\gamma) = 1.$$

Symmetry:
$$B^3_{i,j,k}(u,v,w;\alpha,\beta,\gamma) = B^3_{ijk}(w,v,u;\gamma,\beta,\alpha).$$

For more details on the other properties, please refer to Zhu and Han [44].

Zhu and Han's triangular patches with three parameters α, β, and γ, and control points b_{ijk}, $i+j+k=3$ are defined as

$$P(u,v,w) = \sum_{i+j+k=3} b_{ijk} B^3_{i,j,k}(u,v,w;\alpha,\beta,\gamma), \quad u+v+w=1 \qquad (3)$$

Figure 2 shows the Zhu and Han ordinates, and Figure 3 shows one patch of the Zhu and Han triangular with $\alpha = \beta = \gamma = 3$ (i.e., cubic Bézier triangular).

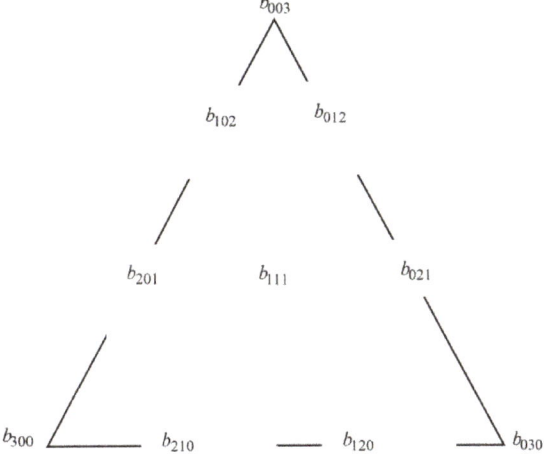

Figure 2. The 10 quartic triangular ordinates (control points).

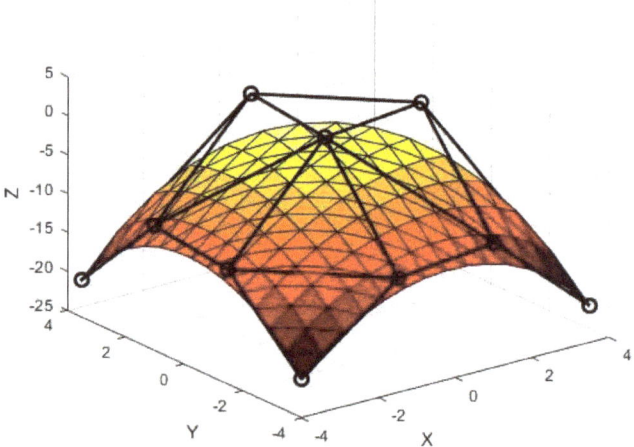

Figure 3. One patch (Zhu and Han [44]).

2.2. Quartic Zhu and Han Triangular Patches

From Equation (2), let $\alpha = \beta = \gamma = 4$, then we obtain the following ten quartic basis functions defined on the triangular domain:

$$
\begin{aligned}
\{B^3_{3,0,0}(u,v,w;\alpha,\beta,\gamma) &= u^4,\ B^3_{0,3,0}(u,v,w;\alpha,\beta,\gamma) = v^4,\ B^3_{0,0,3}(u,v,w;\alpha,\beta,\gamma) \\
&= w^4,\ B^3_{2,1,0}(u,v,w;\alpha,\beta,\gamma) = u^2v\ (3+u),\ B^3_{2,0,1}(u,v,w;\alpha,\beta,\gamma) \\
&= u^2w\ (3+u),\ B^3_{1,2,0}(u,v,w;\alpha,\beta,\gamma) \\
&= uv^2\ (3+v),\ B^3_{0,2,1}(u,v,w;\alpha,\beta,\gamma) \\
&= v^2w\ (3+v),\ B^3_{1,0,2}(u,v,w;\alpha,\beta,\gamma) \\
&= uw^2\ (3+w),\ B^3_{0,1,2}(u,v,w;\alpha,\beta,\gamma) \\
&= vw^2\ (3+w),\ B^3_{1,1,1}(u,v,w;\alpha,\beta,\gamma) = 6uvw.
\end{aligned}
\tag{4}
$$

Figure 4 shows the quartic triangular basis on the triangular domain.

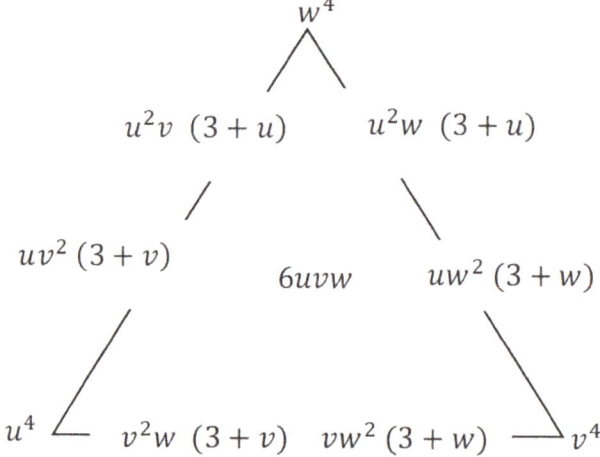

Figure 4. Quartic triangular basis functions.

Thus, the quartic Zhu and Han triangular patch can be defined by

$$P(u,v,w) = u^4 b_{300} + v^4 b_{030} + w^4 b_{003} + u^2 v\,(3+u) b_{210} + (3+u) u^2 w b_{201}$$
$$+(3+v)v^2 u b_{120} + (3+v)v^2 w b_{021} + (3+w)w^2 u b_{102} \qquad (5)$$
$$+(3+w)w^2 v b_{012} + 6 u v w b_{111}$$

The main advantage of Zhu and Han's quartic is that it only requires ten control points to construct one triangular patch; meanwhile, the quartic Bézier triangular will require 15 control points to produce one patch. Furthermore, when the quartic Bézier triangular is used for scattered data interpolation, an optimization method is required to produce the interpolated surface, as discussed in Saaban et al. [35], Piah et al. [36] and Hussain et al. [37,38]. However, if we apply the proposed quartic triangular patches for scattered data interpolation, the optimization is not required since we can employ the cubic precision scheme of Foley and Opitz [30] to construct a C^1 interpolated surface everywhere. So far, this is the first study to apply a a quartic triangular basis but with ten control points for scattered data interpolation.

Figure 5a shows examples of quartic Zhu and Han, and Figure 5b shows the quartic Bézier triangular patch.

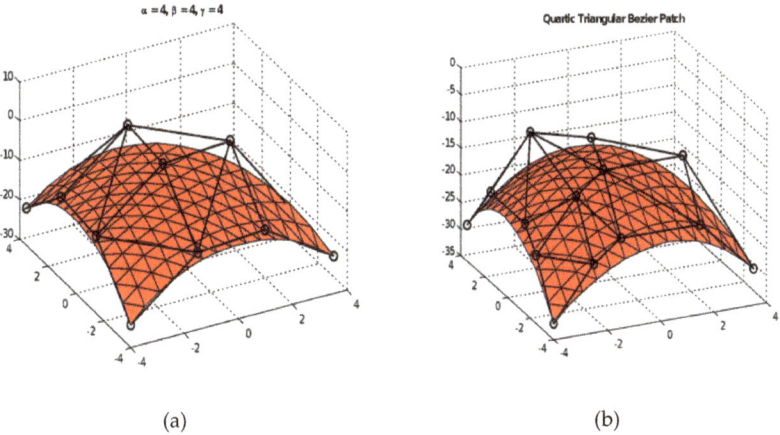

Figure 5. Quartic triangular patches. (a) Quartic Zhu and Han [44]; (b) Quartic Bézier triangular.

2.3. Scattered Data Interpolation Using Quartic Zhu and Han Triangular Patches

To apply the quartic triangular patch defined in Section 2.2 for scattered data, we use the local scheme comprising a convex combination between three local schemes $K_1, K_2,$ and K_3 [1,9,24] such that:

$$P(u,v,w) = \frac{vw K_1 + uw K_2 + uv K_3}{vw + uw + uv}, \quad u+v+w = 1 \qquad (6)$$

The local scheme K_i, $i = 1,2,3$ is obtained by replacing b_{111} in (5) with b^i_{111} to ensure the C^1 condition is satisfied. Given the vertex of the triangle (i.e., $F(V_1) = b_{300}$, $F(V_2) = b_{030}$, and $F(V_3) = b_{003}$), the derivative along e_{jk} (see Figure 6)—that is, the edge connecting two points $(x_j - y_j)$ and $(x_k - y_k)$—is defined as [1,9,24,29]:

$$\frac{\partial P}{\partial e_{jk}} = (x_k - x_j)\frac{\partial F}{\partial x} + (y_k - y_j)\frac{\partial F}{\partial y}$$

Thus

$$b_{210} = F(V_1) + \frac{1}{4}\frac{\partial P}{\partial e_3}(V_1)$$

which can be simplified as

$$b_{210} = b_{300} + \frac{1}{4}\big((x_2 - x_1)F_x(V_1) + (y_2 - y_1)F_y(V_1)\big) \qquad (7)$$

Similarly, the other five ordinates are calculated as follows:

$$b_{201} = b_{300} - \frac{1}{4}\big((x_1 - x_3)F_x(V_1) + (y_1 - y_3)F_y(V_1)\big) \qquad (8)$$

$$b_{021} = b_{030} + \frac{1}{4}\big((x_3 - x_2)F_x(V_2) + (y_3 - y_2)F_y(V_2)\big) \qquad (9)$$

$$b_{120} = b_{030} - \frac{1}{4}\big((x_2 - x_1)F_x(V_2) + (y_2 - y_1)F_y(V_2)\big) \qquad (10)$$

$$b_{102} = b_{003} + \frac{1}{4}\big((x_1 - x_3)F_x(V_3) + (y_1 - y_3)F_y(V_3)\big) \qquad (11)$$

$$b_{012} = b_{003} - \frac{1}{4}\big((x_3 - x_2)F_x(V_3) + (y_3 - y_2)F_y(V_3)\big) \qquad (12)$$

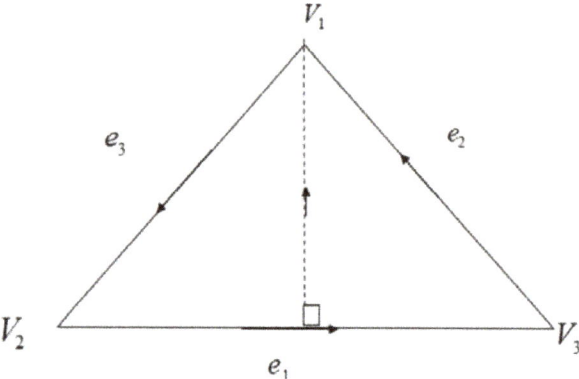

Figure 6. Side-vertex blending.

The remaining b^i_{111}, $i = 1, 2, 3$ is obtained by using the cubic precision of Foley and Opitz [30] as shown in Figure 7. For complete derivation, the reader can refer to [30].

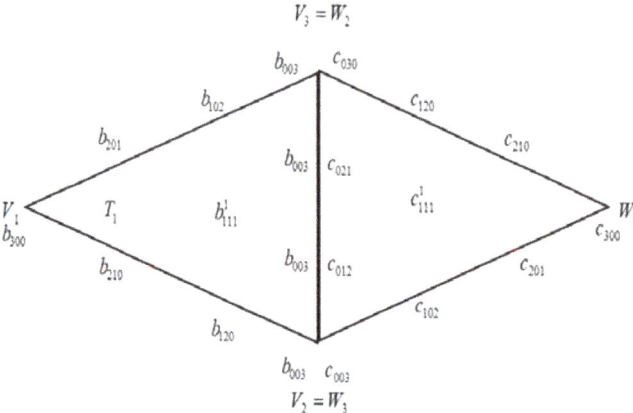

Figure 7. Two adjacent quartic triangular patches.

In order to achieve C^1 continuity along all edges, the following equations must be satisfied:

$$c_{201} = r^2 b_{210} + 2st b_{021} + 2rs b_{120} + s^2 b_{030} + 2rt b_{111}^2 + t^2 b_{012} \tag{13}$$

$$c_{210} = r^2 b_{201} + 2st b_{012} + 2rt b_{102} + s^2 b_{021} + 2rs b_{111}^1 + t^2 b_{003} \tag{14}$$

$$b_{210} = u^2 c_{201} + 2vw c_{012} + 2uw c_{102} + v^2 c_{021} + 2uv c_{111}^1 + w^2 c_{003} \tag{15}$$

$$b_{201} = u^2 c_{210} + 2vw c_{021} + 2uw c_{120} + v^2 c_{030} + 2uv c_{111}^1 + w^2 c_{012} \tag{16}$$

To find c_{111}^1 in (13) and (14), we need to add these equations together. Thus, we obtain

$$c_{111}^1 = \tfrac{1}{2u(v+w)} \big(b_{201} + b_{210} - u^2(c_{210} + c_{201}) - v^2(c_{030} + c_{021}) \big)$$
$$+ w^2(c_{012} + c_{003}) - 2vw(c_{021} + c_{012}) - uvc_{120} - 2uwc_{102}.$$

Similarly, with Equations (15) and (16), we obtain

$$b_{111}^1 = \tfrac{1}{2r(s+t)} \big(c_{201} + c_{210} - r^2(b_{210} + b_{201}) - s^2(b_{030} + b_{021}) \big)$$
$$+ t^2(b_{012} + b_{003}) - 2st(b_{021} + b_{012}) - rsb_{120} - 2rtb_{102}.$$

Now we establish the theorem for the main result.

Theorem 1. *The local scheme defined by (6) is a rational function with degree 7, that is, degree five in numerator and degree two in denominator with C^1 continuity everywhere. It has the following form:*

$$P(u,v,w) = \sum_{\substack{i+j+k=3 \\ i,j,k \neq 1}} b_{ijk} B^3_{i,j,k}(u,v,w) + 6uvw \left(a_1 b_{111}^1 + a_2 b_{111}^2 + a_3 b_{111}^3 \right) \tag{17}$$

with

$$a_1 = \frac{vw}{vw + uw + uv}, \; a_2 = \frac{uw}{vw + uw + uv}, \; a_3 = \frac{uv}{vw + uw + uv} \tag{18}$$

and the barycentric coordinate satisfies $u + v + w = 1$.

The following Algorithm 1 can be used to implement the proposed scheme.

Algorithm 1 (Scattered Data Interpolation)

Step 1: Input scattered data points;
Step 2: Estimate the partial derivative at the data points by using [25];
Step 3: Triangulate the domain of the data points;
Step 4: Calculate the boundary control points using Equations (7)–(12);
Step 5: Calculate inner control points for the local scheme, b_{111}^i, $i = 1, 2, 3$ by using the cubic precision method as in Foley and Opitz [30];
Step 6: Construct the interpolated surface using the convex combination method of three local schemes defined by (6);
Step 7: Calculate CPU time (in seconds), R^2, and maximum error. Repeat steps 1 through 6 for the other scattered data sets.

Below we give the theorem for scattered data interpolation by using quartic Bézier triangular patches.

Theorem 2. *C^1 quartic Bézier triangular patches using minimized sum of squares of principal curvatures* [35].

Let the quartic Bézier triangular patch ($n = 4$) with barycentric coordinates u, v, w be expressed as

$$S(u, v, w) = \sum_{i+j+k=4} c_{ijk} B^4_{ijk}(u, v, w) \quad (19)$$

where $B^4_{ijk}(u, v, w) = \frac{4!}{i!j!k!} u^i v^j w^k$ and c_{ijk} are the Bézier ordinates of S. Let the total number of triangles in the whole triangular mesh be n_t and the total number of interior edges be n_e. $S(x,y)$ which will minimize the functional $I(S(x,y))$ leads to the optimization problem $\mathbf{x}^T Q \mathbf{x} + \mathbf{e}\mathbf{x} + h$, subject to the C^1 continuity constraint:

$$\min {}_x x^T Q x + e x + h \text{ subject to } Ax = b \quad (20)$$

where Q is a $6n_t \times 6n_t$ sparse matrix, $\mathbf{e}$ is a $1 \times 6n_t$ row vector, $\mathbf{x}$ is a $6n_t \times 1$ column vector consisting of unknown ordinates $\left(b^m_{220}, b^m_{211}, b^m_{121}, b^m_{202}, b^m_{112}, b^m_{022}\right)$, $m = 1, \ldots, n_t$, h is a real constant, A is a $3n_e \times 6n_t$ ($3n_e \leq 6n_t$) coefficient matrix, $\mathbf{x}$ is a $6n_t \times 1$ unknown column vector consisting of the remaining ordinates $b_{220}, b_{202}, b_{022}, b_{211}, b_{121}$ and b_{112} to be determined for the entire triangular mesh, and $\mathbf{b}$ is a $3n_e \times 1$ constant column vector. The optimization problem stated in (18) was solved using the optimization toolbox in MATLAB 2017 on Intel® Core ™ i5-8250U 1.60 GHz.

Note that the optimization problem in Theorem 2 is obtained by using a minimized principal curvature norm with respect to the C^1 continuity constraint, which results in a global method for scattered data interpolation. Meanwhile, by using the proposed scheme in this study, the resulting surface is local.

3. Results and Discussion for Scattered Data Interpolation

We tested the proposed scheme using two well-known test functions $F_1(x, y)$ and $F_2(x, y)$:

$$F_1(x, y) = 0.75 e^{(-(9x-2)^2 + (9y-2)^2/4)} + 0.75 e^{(-(9x+1)^2/49 - (9y+1)^2/10)} \\ + 0.5 e^{(-(9x-7)^2 + (9y-3)^2/4)} - 0.2 e^{(-(9x-4)^2 - (9y-7)^2)} \quad (21)$$

$$F_2(x, y) = \frac{(1.25 + \cos \cos (5.4y))}{6 + 6(3x - 1)^2} \quad (22)$$

We implemented the proposed scheme using MATLAB 2017 version on Intel® Core ™ i5-8250U 1.60 GHz. MATLAB coding was developed based on Algorithm 1. About 25 MATLAB functions were used to obtain all the results.

We chose 36 data point samples in the domain $[0, 1] \times [0, 1]$, as shown in Table 1. Figure 8 shows the Delaunay triangulation for the data. Figure 6 shows examples of surface interpolation for both functions. Comparing Figures 9 and 10, the surfaces produced by the proposed scheme visually look smoother than the surfaces obtained from the quartic Bézier triangular of Saaban et al. [35] and Piah et al. [36]. Figure 10 shows an example of scattered data interpolation using quartic Bézier triangular patches.

To validate the proposed scheme, we calculated the maximum error (Max Error) and coefficient of determination (COD; i.e., R^2) for both functions and compared them with those obtained for quartic Bézier triangular for three different numbers of points i.e., 100, 65, and 36 for both functions $F_1(x, y)$ and $F_2(x, y)$. Functions 1 and 2 represent $F_1(x, y)$ and $F_2(x, y)$, respectively.

Table 2 shows the error analysis for both tested functions by using (a) quartic Zhu and Han and (b) quartic Bézier triangular. Meanwhile, Table 3 shows CPU time in seconds. From Table 2, we can see that the proposed quartic triangular patches for scattered data interpolation gave smaller Max Error values than the quartic Bézier triangular. Additionally, the proposed scheme gave higher R^2 values for all numbers of data points (100, 65, and 36). From Table 3, the proposed scheme required less CPU time than the quartic Bézier. For instance, for 100 data points, the proposed scheme only required 0.71 s for data from function $F_1(x, y)$ and 0.42 s for data from function $F_2(x, y)$, compared with the quartic Bézier which requires 5.6 s and 3.57 s for 100 data points from functions $F_1(x,y)$ and $F_2(x,y)$, respectively. Thus, the proposed scheme in this study gave very good results, and was better at treating scattered data than using the quartic Bézier triangular proposed by Piah et al. [36], Saaban et al. [35], and Hussain et al. [37,38]. We conclude that the proposed scheme required less CPU time than the quartic Bézier triangular. This reduction of CPU time consumption is an advantage when the goal is to construct a surface with thousands of data points or big data.

Table 1. Data points.

x	y	$F_1(x,y)$	$F_2(x,y)$	x	y	$F_1(x,y)$	$F_2(x,y)$
0	0	0.7664	1.3333	0.80	0.85	0.0823	1.2431
0.50	0	0.4349	1.3833	0.85	0.65	0.1412	1.2043
1.00	0	0.1076	1.2833	1.00	0.50	0.1610	1.2199
0.15	0.15	1.1370	1.3382	1.00	1.00	0.0359	1.2712
0.70	0.15	0.4304	1.3020	0.50	1.00	0.1460	1.3346
0.50	0.20	0.5345	1.3128	0.10	0.85	0.2935	1.2363
0.25	0.30	1.0726	1.2423	0	1.00	0.2703	1.3029
0.40	0.30	0.7134	1.2421	0.25	0	0.8189	1.4069
0.75	0.40	0.5903	1.2139	0.75	0	0.2521	1.3150
0.85	0.25	0.5088	1.2607	0.25	1.00	0.2222	1.3496
0.55	0.45	0.3823	1.1613	0	0.25	0.8026	1.2683
0	0.50	0.4818	1.1747	0.75	1.00	0.0810	1.2913
0.20	0.45	0.6458	1.1412	0	0.75	0.3395	1.1987
0.45	0.55	0.2946	1.1037	1.00	0.25	0.2302	1.2573
0.60	0.65	0.1920	1.1552	1.00	0.75	0.0504	1.2295
0.25	0.70	0.2930	1.1240	0.19	0.19	1.2118	1.3229
0.40	0.80	0.0515	1.1887	0.32	0.75	0.2029	1.1477
0.65	0.75	0.1372	1.1961	0.79	0.46	0.4777	1.2041

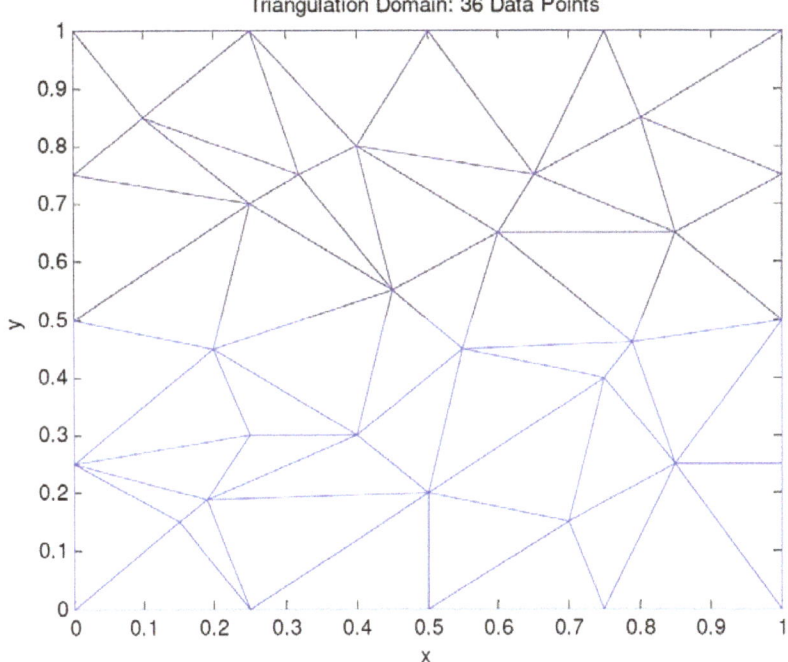

Figure 8. Delaunay triangulation of data in Table 1.

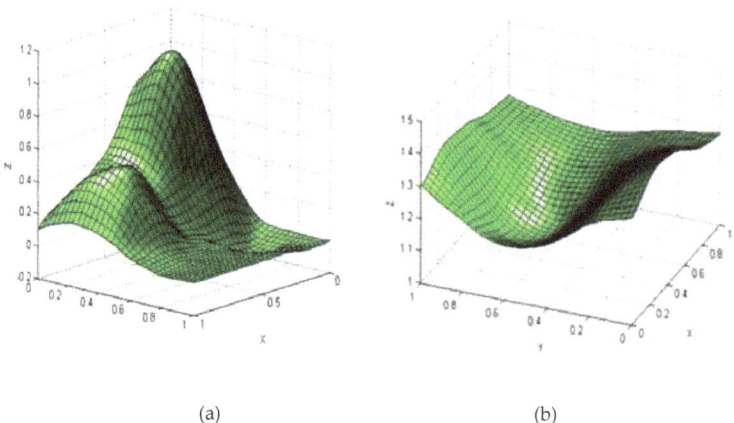

Figure 9. Surface reconstruction using the proposed scheme. (**a**) For $F_1(x,y)$; (**b**) For $F_2(x,y)$.

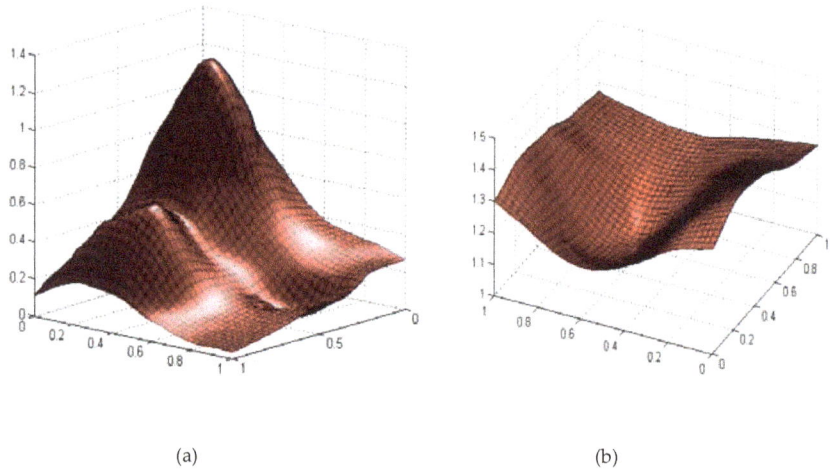

Figure 10. Surface reconstruction using quartic Bézier triangular. (**a**) For $F_1(x,y)$; (**b**) For $F_2(x,y)$.

Table 2. Error analysis.

Num. of Data Points	Function	Max Error		R^2	
		The Proposed Scheme	Quartic Bézier [35]	The Proposed Scheme	Quartic Bézier [35]
100	1	3.436×10^{-2}	3.598×10^{-2}	0.99936	0.99934
	2	4.500×10^{-2}	7.61×10^{-2}	0.99977	0.99967
65	1	6.410×10^{-2}	6.586×10^{-2}	0.99720	0.99733
	2	1.732×10^{-2}	1.562×10^{-2}	0.99796	0.99793
36	1	9.740×10^{-2}	9.973×10^{-2}	0.99211	0.99256
	2	2.675×10^{-2}	2.762×10^{-2}	0.99332	0.99208

This can be seen clearly from Table 3. With the largest number of data points, the CPU time for the proposed scheme was approximately 12.5% that of the CPU time required for the quartic Bézier triangular based scheme. This is very significant, especially when the user wants to render and reconstruct surfaces obtained from very dense data sets. Many studies in scattered data interpolation usually involve the use of Shepard-type interpolants such as Shepard triangular schemes for scattered data interpolation [6,11–13]. We also implemented the Shepard triangular to the same data sets as listed in Table 1. Tables 4 and 5 show the error analysis for the schemes of Cavoretto et al. [6], Dell'Accio et al. [12,13], and Dell'Accio and Di Tommaso [11]. Based on CPU time, for all tested data sets, the proposed scheme was faster than the schemes in [6,11–13], except for the case with 100 data points for the data from function $F_1(x,y)$. Considering the Max Error, the proposed scheme was better than all four schemes except for the case with 100 data points from function $F_2(x,y)$. Therefore, we can conclude that the proposed scheme is better than quartic triangular patch and the Shepard triangular based schemes [6,11–13].

Table 3. CPU time (in seconds).

Num. of Data Points	Function	CPU Time (in Seconds)	
		The Proposed Scheme	Quartic Bézier [35]
100	1	0.7097807844	5.6002481345
	2	0.4234289196	3.5703151686
65	1	0.2741610887	1.5474957467
	2	0.2363209917	1.3271791002
36	1	0.1298699059	0.5886074910
	2	0.1163838547	0.4703613961

Table 4. Error analysis.

Num. of Data Points	Function	Max Err				
		Dell'Accio et al. [12]	Dell'Accio and Di Tommaso [11]	Dell'Accio et al., [12] and Cavoretto et al. [6]	Dell'Accio et al. [12]	Dell'Accio et al. [13] and Cavoretto et al. [6]
100	1	5.2990×10^{-2}	8.6648×10^{-2}	1.0970×10^{-1}	6.2438×10^{-2}	5.3936×10^{2}
	2	1.8617×10^{-2}	5.0590×10^{-2}	3.2842×10^{-2}	1.5449×10^{-2}	1.9619×10^{-2}
65	1	1.0147×10^{-1}	1.1864×10^{-1}	1.1221×10^{-1}	7.6266×10^{-2}	7.1704×10^{-2}
	2	6.4329×10^{-2}	3.7704×10^{-2}	3.5962×10^{-2}	2.7322×10^{-2}	2.8894×10^{-2}
36	1	1.2822×10^{-1}	1.6219×10^{-1}	1.3564×10^{-1}	1.1371×10^{-1}	9.8914×10^{-2}
	2	7.9686×10^{-2}	5.6713×10^{-2}	5.3611×10^{-2}	5.1806×10^{-2}	4.6253×10^{-2}

Table 5. CPU time (in seconds).

Num. of Data Points	Function	CPU Time (Second)				
		Dell'Accio et al. [12]	Dell'Accio and Di Tommaso [11]	Dell'Accio et al., [12] and Cavoretto et al. [6]	Dell'Accio et al. [12]	Dell'Accio et al. [13] and Cavoretto et al. [6]
100	1	0.490380	1.894685	0.480296	0.423177	0.401825
	2	0.502923	1.881019	0.501286	0.428949	0.417658
65	1	0.486381	1.882420	0.478175	0.424583	0.424086
	2	0.484984	1.849458	0.459517	0.424870	0.397832
36	1	0.437090	1.707668	0.445523	0.415162	0.445566
	2	0.448242	1.686242	0.448589	0.417875	0.421869

4. Positivity-Preserving Scattered Data Interpolation

In this section, we apply the proposed scheme discussed in the previous section to preserve the positivity of scattered data sets. To do this, first we derive the sufficient condition for the positivity of the quartic triangular spline defined in (5). Finally, the rational corrected scheme defined by (19) will be used to construct a positive surface with C^1 continuity.

To derive the sufficient condition for the positivity of the quartic spline triangular patch, we adopted a similar approach to Saaban et al. [35] and Piah et al. [36]. Assume that the quartic ordinates at the vertices are strictly positive such $b_{300}, b_{030}, b_{003} > 0$. Let $A = b_{300}, B = b_{030}, C = b_{003}$, therefore

$A, B, C > 0$ (see Figure 11). Meanwhile, let the other ordinates have the same value, that is $-t < 0$ where $t > 0$. Thus, Equation (5) becomes

$$\begin{aligned} P(u,v,w) &= Au^4 + Bv^4 + Cw^4 \\ &\quad -t\big(u^2v\,(3+u) + (3+u)u^2w + (3+v)v^2u + (3+v)v^2w + (3+w)w^2u \\ &\quad + (3+w)w^2v + 6uvw\big) \\ &= Au^4 + Bv^4 + Cw^4 - t\big(1 - u^4 - v^4 - w^4\big) \\ &= (A+t)u^4 + (B+t)v^4 + (C+t)w^4 - t \end{aligned} \quad (23)$$

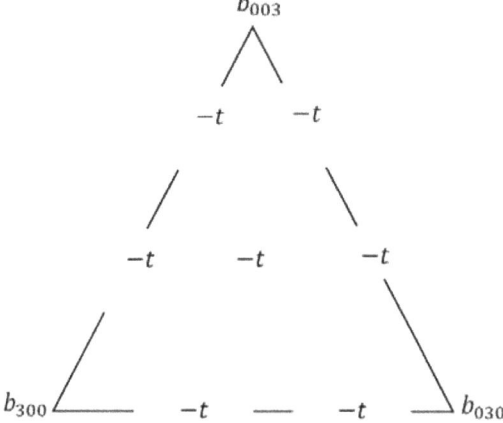

Figure 11. Quartic triangular ordinates arrangement for positivity preservation.

From (23) we can observe that when $t = 0$ then $P(u,v,w) > 0$. Meanwhile as t increases, $P(u,v,w)$ decreases. We want to find the value of $t = t_0$ when the minimum value of $P(u,v,w) = 0$. By taking first partial derivatives, we will obtain the following:

$$\begin{aligned} \frac{\partial P(u,v,w)}{\partial u} &= 4(A+t)u^3, \\ \frac{\partial P(u,v,w)}{\partial v} &= 4(B+t)v^3, \\ \frac{\partial P(u,v,w)}{\partial w} &= 4(C+t)w^3. \end{aligned} \quad (24)$$

The minimum value of $P(u,v,w)$ occurs when

$\frac{\partial P}{\partial u} - \frac{\partial P}{\partial v} = 0$ and $\frac{\partial P}{\partial u} - \frac{\partial P}{\partial w} = 0$ or equivalently.

$$\frac{\partial P}{\partial u} = \frac{\partial P}{\partial v} = \frac{\partial P}{\partial w} \quad (25)$$

From Equation (25) we have
$\frac{u^3}{v^3} = \frac{B+t}{A+t}$ and $\frac{u^3}{w^3} = \frac{C+t}{A+t}$
Hence:

$$u^3 : v^3 : w^3 = \frac{1}{A+t} : \frac{1}{B+t} : \frac{1}{C+t}.$$

Since $u + v + w = 1$, we obtain the following relations:

$$u = \frac{\frac{1}{(A+t)^{1/3}}}{\frac{1}{(A+t)^{1/3}} + \frac{1}{(B+t)^{1/3}} + \frac{1}{(C+t)^{1/3}}},$$

$$v = \frac{\frac{1}{(B+t)^{1/3}}}{\frac{1}{(A+t)^{1/3}} + \frac{1}{(B+t)^{1/3}} + \frac{1}{(C+t)^{1/3}}}, \text{ and}$$

$$w = \frac{\frac{1}{(C+t)^{1/3}}}{\frac{1}{(A+t)^{1/3}} + \frac{1}{(B+t)^{1/3}} + \frac{1}{(C+t)^{1/3}}}.$$

Substituting this value into (23) we obtain the minimum value of $P(u,v,w)$ i.e., $P(u,v,w)\frac{1}{\left[\frac{1}{(A+t)^{1/3}} + \frac{1}{(B+t)^{1/3}} + \frac{1}{(C+t)^{1/3}}\right]^3}_{min}$ which can be simplified to

$$P(u,v,w) \frac{t}{\left[\frac{1}{(A/t+1)^{1/3}} + \frac{1}{(B/t+1)^{1/3}} + \frac{1}{(C/t+1)^{1/3}}\right]^3}_{min} \tag{26}$$

Now, $P(u,v,w)_{min}$ when

$$\frac{1}{(A/t+1)^{1/3}} + \frac{1}{(B/t+1)^{1/3}} + \frac{1}{(C/t+1)^{1/3}} = 1 \tag{27}$$

Let $s = 1/t$, then

$$G(s) = \frac{1}{(As+1)^{1/3}} + \frac{1}{(Bs+1)^{1/3}} + \frac{1}{(Cs+1)^{1/3}} \tag{28}$$

Then Equation (27) can be written as $G(s) = 1, s \geq 0$. This equation can be solved by using regula-falsi method with suitable choice of initial guess.

Since $A, B, C > 0$ and $s \geq 0$ then $G'(s) < 0$ and $G''(s) > 0$. (see Figure 12). Thus, the curve is convex on that region. Let $X = max(A,B,C)$ and $Y = min(A,B,C)$, then the following holds

$$\frac{3}{\sqrt[3]{Xs+1}} \leq G(s) \leq \frac{3}{\sqrt[3]{Ys+1}}$$

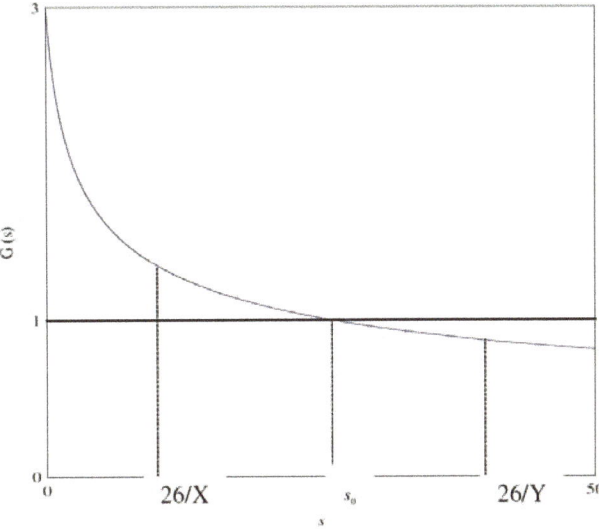

Figure 12. Function $G(s)$ for $s \geq 0$.

Such that
$$G\left(\frac{26}{X}\right) \geq 1 \text{ and } G\left(\frac{26}{Y}\right) \leq 1.$$

Figure 12 shows the example of the relative locations of $\frac{26}{X}$ and $\frac{26}{Y}$ and s_0.

Now we establish the main theorem for positivity preservation using the proposed scheme.

Theorem 3. *Consider the quartic triangular patch $P(u,v,w)$ with vertex $A = b_{300}, B = b_{030}, C = b_{003}$, such that $A, B, C > 0$. If the remaining quartic triangular ordinates are equal to $-t_0$ where $t_0 = \frac{1}{s_0}$ is a unique solution to (28), then $P(u,v,w) \geq 0$ for all $u, v, w \geq 0$ and $u + v + w = 1$.*

Some observations from Theorem 3 can be made as follows:
Let $A = B > C = 1$. Then, we have
$$G(s) = \frac{2}{(As+1)^{1/3}} + \frac{1}{(s+1)^{1/3}}$$

Therefore, as $A \to \infty$, then $G(s) \to \frac{1}{(s+1)^{1/3}}$. Hence, $s_0 \to 0$ and therefore $t_0 \to \infty$. Thus, the ordinate values are unbounded compared with the work of Chan and Ong [7] in which the Bézier ordinates are bounded by a lower bound $-1/3$.

Remark 1. *The sufficient condition for the positivity of the quartic triangular patch developed in this study is the same as the sufficient condition for the quartic Bézier triangular patch developed in Saaban et al. [35]. The main difference is that the proposed quartic polynomial only requires ten control points (or ordinates) as compared to the quartic Bézier triangular which requires 15 control points and involves some optimization problems as shown in Saaban et al. [35] and Hussain et al. [37,38]. Therefore, the proposed positivity preservation using a quartic triangular patch requires less computation time than some established schemes for scattered data interpolation.*

The final construction of the positive scattered surface is described below:
1. Input positive scattered data points;
2. Triangulate the scattered data using Delaunay triangulation;
3. Assign the first partial derivative at the respective data sites and adjust if necessary, to provide the positivity preservation;
4. The C^1 triangular surface is constructed via convex combination between three local schemes;
5. Repeat Steps 1 through 4 for other positive scattered data sets.

5. Numerical Results and Discussion for Positivity-Preserving Scattered Data Interpolation

After we derive the sufficient condition for the positivity of quartic triangular patch, the final C^1 scattered data scheme for positivity preservation can be written as follows:

$$P(u,v,w) = \sum_{\substack{i+j+k=3 \\ i,j,k \neq 1}} b_{ijk} B^3_{i,j,k}(u,v,w) + 6uvw\left(a_1 b^1_{111} + a_2 b^2_{111} + a_3 b^3_{111}\right) \quad (29)$$

with
$$a_1 = \frac{vw}{vw+uw+uv}, a_2 = \frac{uw}{vw+uw+uv}, a_3 = \frac{uv}{vw+uw+uv} \quad (30)$$

We test the proposed scheme by using four well-known test functions given below:

$$F_1(x,y) = \begin{cases} 1.0 & \text{if}(y-x) \geq 0.5 \\ 2(y-x) & \text{if}\, 0.5 \geq (y-x) \geq 0.0 \\ \dfrac{\cos(4\pi\sqrt{(x-1.5)^2+(y-0.5)^2}+1}{2} & \text{if}\,(x-1.5)^2+(y-0.5)^2 \leq \tfrac{1}{16} \\ 0 & \text{elsewhere on}\,[0,2]x[0,1] \end{cases}$$

$$F_2(x,y) = 1.025 - 0.75e^{-(6x-1)^2-(6y-1)^2} - 0.75e^{-(9x+1)^2/49-(9y+1)^2/10} \\ -0.5e^{-(9x-7)^2-(9y-3)^2} - 0.5e^{-(10x-4)^2-(10y-7)^2}$$

$$F_3(x,y) = x^4 + y^4$$

$$F_4(x,y) = e^{-(5-10x)^2/2} + 0.75e^{-(5-10y)^2/2} + 0.75e^{-(5-10x)^2/2}e^{-(5-10y)^2/2}$$

The positive test functions F_1, F_2, F_3, and F_4 were evaluated on 36, 33, 26, and 100 node points respectively (Tables 6–9) where all function values were greater than or equal to zero. The nodes of 36 and 33 points were defined on a rectangular domain (Figure 13a,b), while the 26- and 100-point nodes were defined on a sparse non-rectangular domain (Figure 13c,d). Tables 8 and 9 show examples of irregular scattered data sets.

Table 6. Value of F_1 on 36 node points.

x	y	$F_1(x,y)$	x	y	$F_1(x,y)$	x	y	$F_1(x,y)$
0	0	0	0.35	0	0	1.4	0.8	0
0.2	0.2	0	0.8	0	0	1.65	0.75	0
0.5	0.2	0	0.1	0.85	1	2	1	0
0.4	0.4	0	0	0.25	0.5	1.25	0	0
0.75	0.35	0	0.8	1	0.4	1.7	0	0
0	0.5	1	2	0	0	1.25	1	0
0.25	0.5	0.5	1.4	0.3	0.0272	1.7	1	0
0.25	0.75	1	1.75	0.45	0	2	0.35	0
0.55	0.75	0.4	1.2	0.45	0	2	0.7	0
0.7	0.6	0	1.45	0.5	0.9045	1.05	0.2	0
0.5	1	1	1.6	0.3	0.0272	1	0.5	0
0	1	1	1.25	0.7	0	0.95	0.8	0

Table 7. Value of F_2 on 33 node points.

x	y	$F_2(x,y)$	x	y	$F_2(x,y)$	x	y	$F_2(x,y)$
0	0	0.2586	0	0.50	0.5960	0.50	1.00	0.8762
0.50	0	0.6429	0.25	0.45	0.6264	0.10	0.85	0.7316
1.00	0	0.9174	0.45	0.55	0.7981	0	1.00	0.7547
0.25	0.20	0.0056	0.60	0.65	0.8336	0.25	0	0.2629
0.70	0.15	0.6012	0.25	0.70	0.7862	0.75	0	0.7739
0.50	0.20	0.6329	0.40	0.80	0.9941	0.25	1.00	0.8026
0.30	0.30	0.4199	0.65	0.75	0.8825	0	0.25	0.2792
0.45	0.35	0.6618	0.80	0.85	0.9427	0.75	1.00	0.9440
0.75	0.40	0.4361	0.85	0.65	0.8838	0	0.75	0.6865
0.85	0.25	0.5164	1.00	0.50	0.8640	1.00	0.25	0.7948
0.55	0.45	0.6724	1.00	1.00	0.9891	1.00	0.75	0.9746

Table 8. Value of F_3 on 26 node points.

x	y	$F_3(x,y)$	x	y	$F_3(x,y)$	x	y	$F_3(x,y)$
0.9375	−0.4063	0.7997	0.0469	−0.7656	0.3436	−0.5156	−0.1094	0.0708
−0.1719	1.0000	1.0009	−0.7813	−0.8906	1.0017	0.4844	0.1406	0.0554
−0.8906	−0.0938	0.6292	0.0625	0.3750	0.0198	−0.4531	0.1563	0.0427
−0.0625	−0.6719	0.2038	−0.7656	−0.2969	0.3513	0.7031	0.3281	0.2560
−0.8750	−0.6250	0.7388	0.0938	0.1250	0.0003	−0.4219	0.4688	0.0800
0	0.7969	0.4033	−0.6875	0.3750	0.2432	0.9063	−0.5938	0.7990
−0.8438	−0.5313	0.5866	0.1094	0.3281	0.0117	−0.2344	0.1406	0.0034
0.0313	0.5313	0.0797	−0.5625	−0.6563	0.2856	0.9688	0.7188	1.1476
−0.8438	0.1563	0.5075	0.1563	0.4531	0.0427			

Table 9. Value of F_4 on 100 node points.

x	y	$F_4(x,y)$	x	y	$F_4(x,y)$	x	y	$F_4(x,y)$
0.0096	0.3083	0.119425	0.3307	0.5159	1.155816	0.6677	0.6764	0.442125
0.0216	0.245	0.029055	0.3379	0.9426	0.268844	0.6814	0.8444	0.195332
0.0298	0.8614	0.00111	0.3439	0.48	1.248258	0.6888	0.3273	0.365476
0.0417	0.0978	0.000258	0.353	0.1783	0.345127	0.6941	0.1894	0.158965
0.047	0.3648	0.300748	0.3636	0.1147	0.395081	0.7062	0.0646	0.119387
0.0563	0.7156	0.073456	0.3766	0.8226	0.473071	0.7161	0.018	0.096822
0.0647	0.5311	0.714724	0.3822	0.2271	0.526808	0.7317	0.8905	0.068664
0.074	0.9756	0.000124	0.387	0.4074	1.274589	0.7371	0.4161	0.619367
0.0874	0.1781	0.004419	0.3973	0.8875	0.590813	0.7462	0.4689	0.797375
0.0935	0.5453	0.677296	0.4171	0.7632	0.749339	0.7567	0.2175	0.051462
0.1032	0.1604	0.00273	0.4256	0.9973	0.758236	0.77	0.5734	0.613977
0.111	0.7837	0.013932	0.4299	0.496	2.117705	0.7879	0.8853	0.016309
0.1181	0.9982	0.000684	0.4373	0.341	1.207499	0.7944	0.8018	0.021117
0.1252	0.6911	0.121795	0.4705	0.2498	1.021601	0.8164	0.6389	0.294451
0.1327	0.105	0.001483	0.4737	0.6409	1.512454	0.8193	0.8931	0.006444
0.144	0.8185	0.00648	0.4879	0.1059	0.99334	0.8368	0.1001	0.003695
0.1565	0.7086	0.088121	0.494	0.5412	2.374907	0.8501	0.279	0.067559
0.1651	0.4457	0.653239	0.5055	0.009	0.998497	0.8588	0.9083	0.001782
0.1786	0.1178	0.006221	0.5163	0.8784	0.987962	0.8646	0.3259	0.166279
0.1886	0.3189	0.154486	0.5219	0.5516	2.273778	0.8792	0.8319	0.003798
0.2017	0.9668	0.011703	0.5349	0.4039	1.858252	0.8838	0.0509	0.000664
0.21	0.7572	0.042784	0.5483	0.1654	0.895154	0.89	0.9708	0.000509
0.2147	0.2017	0.025997	0.557	0.2965	1.02504	0.897	0.5121	0.745189
0.2204	0.3232	0.180361	0.5639	0.366	1.370095	0.9045	0.286	0.076266
0.2344	0.4369	0.662063	0.5785	0.0367	0.734861	0.9084	0.9582	0.00026
0.241	0.8908	0.035317	0.5864	0.9502	0.688545	0.9204	0.6183	0.372734
0.2528	0.0647	0.047165	0.5929	0.2638	0.725544	0.9348	0.378	0.356442
0.2571	0.5693	0.673111	0.5988	0.9277	0.613938	0.9435	0.401	0.459525
0.2733	0.2947	0.174707	0.6118	0.5378	1.60735	0.949	0.9479	7.49E−05
0.2854	0.4332	0.760009	0.6252	0.7375	0.521789	0.957	0.7425	0.039667
0.2902	0.3347	0.323199	0.6331	0.4675	1.417194	0.9772	0.8883	0.00041
0.2965	0.7436	0.169566	0.6399	0.9186	0.375998	0.9983	0.5497	0.66287
0.302	0.1066	0.141203	0.6489	0.0417	0.330061			
0.3126	0.8845	0.173287	0.6559	0.1291	0.29764			

For test function F_1 as the data in Table 6, the interpolated surface did not preserve the positivity of the original surface for the C^1 Zhu and Han quartic (from Theorem 1), as shown in Figure 14a with calculated $\min_{(x,y)\in D} F_1(x,y) = -0.039975$. Observe that these surfaces cross the xy-plane at a number of places. After applying positivity-preserving methods from Theorem 3, the result is shown in Figure 14b, where the interpolated surfaces lie above or on the xy-plane $\min_{(x,y)\in D} F_1(x,y) = 0$.

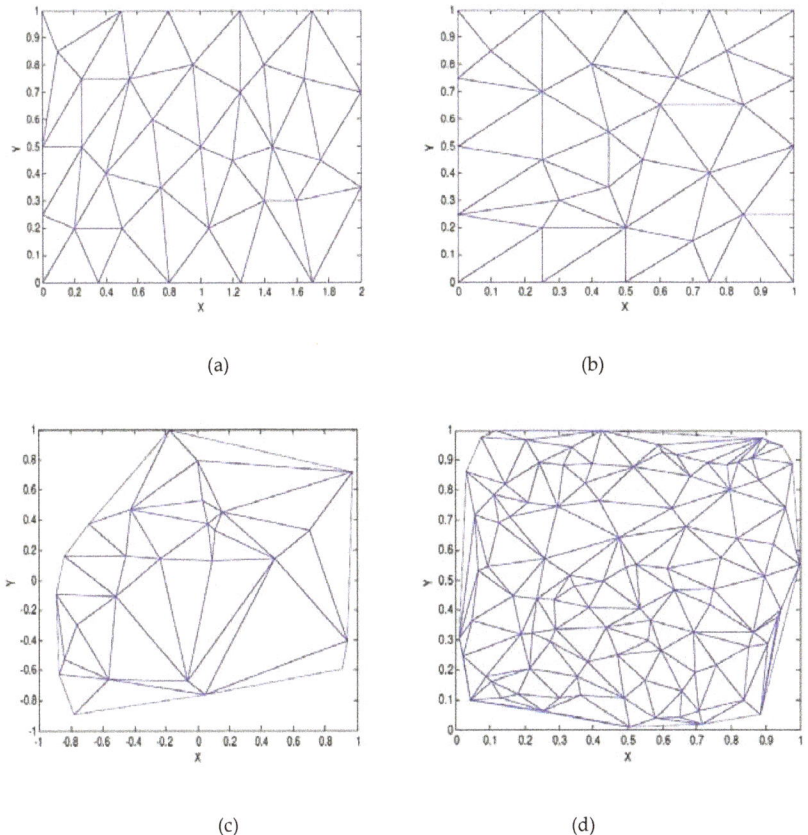

Figure 13. Triangulation domain using Delaunay triangulation: (**a**) 36 node points; (**b**) 63 node points; (**c**) 26 node points; (**d**) 100 node points.

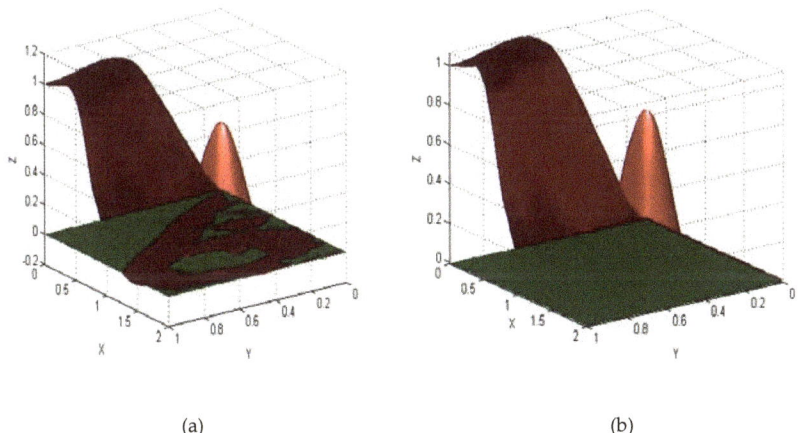

Figure 14. C^1 quartic Zhu and Han interpolated surface (data in Table 6): (**a**) without positivity preserved; (**b**) with positivity preserved from Theorem 3.

For test function F_2 as the data in Table 7, the interpolated surface did not preserve the positivity of the original surface for the C^1 Zhu and Han quartic as shown in Figure 15a, with calculated $\min_{(x,y)\in D} F_2(x,y) = -0.039975$. These surfaces cross the xy-plane at a number of places. Using the proposed positivity-preserving methods, the interpolated surface lies above or on the xy-plane, as shown in Figure 15b, with calculated $\min_{(x,y)\in D} F_2(x,y) = 0.0072657$.

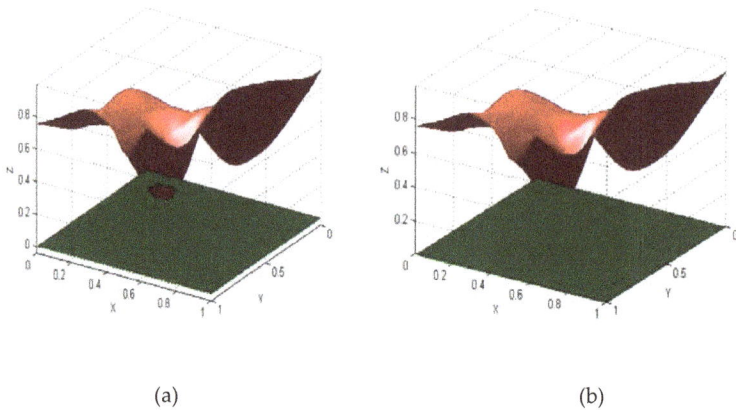

(a)   (b)

Figure 15. C^1 quartic Zhu and Han interpolated surface (data Table 7): (**a**) without positivity preserved; (**b**) with positivity preserved from Theorem 3.

For the third test function defined on a sparse non-rectangular domain (data in Table 8), the interpolated surface did not preserve the positivity, as shown in Figure 16a where the surface crosses below the xy-plane with $\min_{(x,y)\in D} F_3(x,y) = -0.0053288$ and the positivity-preserving interpolated surface using the proposed scheme is shown in Figure 16b where the surface lies above or on the xy-plane, with calculated.

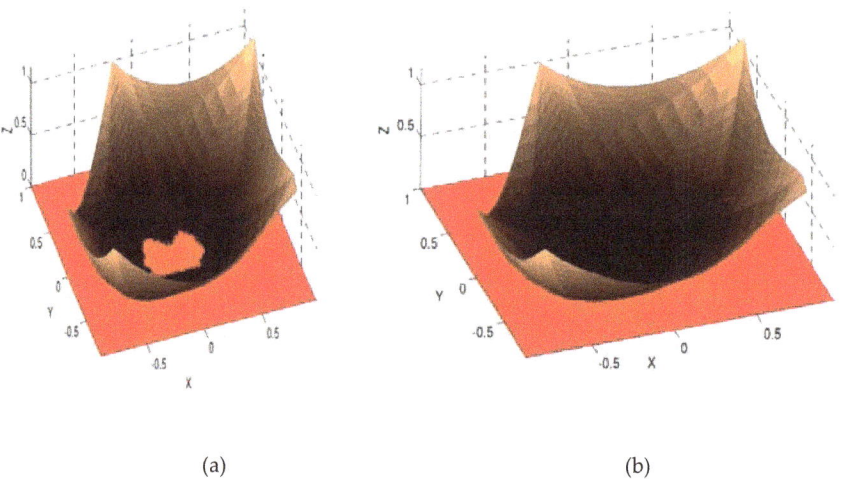

(a)   (b)

Figure 16. C^1 quartic Zhu and Han interpolated surface (data Table 8): (**a**) without positivity preserved; (**b**) with positivity preserved from Theorem 3.

The interpolated surface of the Zhu and Han C^1 quartic without positivity preservation is given in Figure 17a, with calculated $\min_{(x,y) \in D} F_4(x, y) = -0.67634$, while the positivity-preserved surface lying above the xy-plane is illustrated in Figure 17b with calculated $\min_{(x,y) \in D} F_4(x, y) = 0.000074928$.

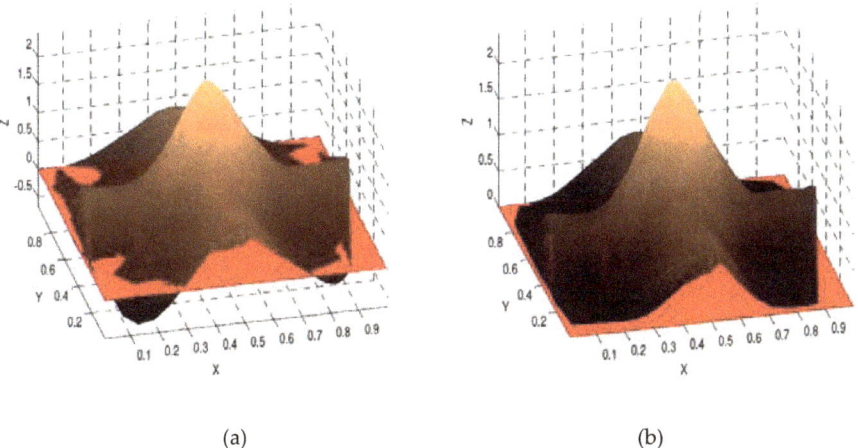

(a) (b)

Figure 17. C^1 quartic Zhu and Han interpolated surface (data Table 9): (**a**) without positivity preserved; (**b**) with positivity preserved from Theorem 3.

We also calculated the CPU time (in seconds), maximum error, and coefficient of determination (R^2) for the positivity-preserving scattered data interpolation as shown in Tables 10 and 11. Once again, the proposed scheme was superior to the quartic Bézier triangular patch. For positivity preservation in scattered data interpolation with dense data sets (i.e., 100 data points with 1697 points of evaluation), the proposed scheme only required 0.5168 s, compared with the quartic Bézier which required 18.5996 s. This is about 36 times faster than the times obtained by the schemes of Saaban et al. [35] and Piah et al. [36]. Roughly, the proposed scheme only required about 2.78% of the CPU times of schemes [35,36]. This is very significant when we want to visualize thousands of scattered data points.

Table 10. CPU time (in seconds).

Size Data	Interpolation Points	Function	CPU Time (in Seconds)	
			The Proposed Scheme	Quartic Bézier [35]
36	1296	F_1	0.6587	1.5653
33	1296	F_2	1.0159	2.6322
25	377	F_3	0.0935	1.1535
100	1697	F_4	0.5168	18.5996

Table 11. Error analysis.

Number of Evaluation Points	Function	Max Error		R^2	
		The proposed scheme	Quartic Bézier [35]	The proposed scheme	Quartic Bézier [35]
1296	F_1	0.2729197868113	0.282452475456	0.9800905	0.9760174674571
1296	F_2	0.6346772647732	0.633667822156	0.819140703	0.8244631293045
377	F_3	0.2716512377297	0.311611498819	0.93773671136	0.935885979498
1697	F_4	0.64080639264362	0.4808917225171	0.98734351099	0.988298717425

Our final example is devoted to the coronavirus disease 2019 (COVID-19) cases at Selangor State and Klang Valley in Malaysia until 15 April 2020. There were 5072 positive cases in Malaysia on

15 April 2020. Selangor and Klang Valley alone had about 2296 positive cases. This represents 45.27% of all COVID-19 cases in Malaysia. Table 12 shows the number of positive COVID-19 cases in 14 districts of Selangor and Klang Valley, including Putrajaya [45].

Table 12. Coronavirus disease 2019 (COVID-19) cases at Selangor and Klang Valley in Malaysia until 15 April 2020.

Label	District	Longitude	Latitude	COVID-19 Cases
A	Hulu Langat	101.7620249	3.0727692	440
B	Petaling	101.664208	3.086134	363
C	Klang	101.449611	3.043125	171
D	Gombak	101.714574	3.233044	142
E	Sepang	101.709401	2.800862	68
F	Hulu Selangor	101.641482	3.52361	49
G	Lembah Pantai	101.672189	3.104444	577
H	Kuala Selangor	101.34555	3.362102	35
I	Kuala Langat	101.496182	2.836562	25
J	Sabak Bernam	101.058059	3.687115	23
K	Kepong	101.623581	3.2059	142
L	Titi Wangsa	101.695278	3.173573	129
M	Cheras	101.71649	3.107178	78
N	Putrajaya	101.684046	2.918	54

Figures 18 and 19 show the example of surface interpolation for COVID-19 scattered data listed in Table 12. Figure 18 shows the interpolated surface without positivity preservation. Figure 19 shows the interpolated surface after we applied the positivity-preserving scheme. Clearly, Figure 19 is suitable for the relevant agency to visualize the number of COVID-19 cases. Then, they can prepare any contingency plan for the spread of COVID-19. They could also try to minimize the spread of COVID-19. This is very crucial, since at the time of writing there are no available vaccines to cure the patients.

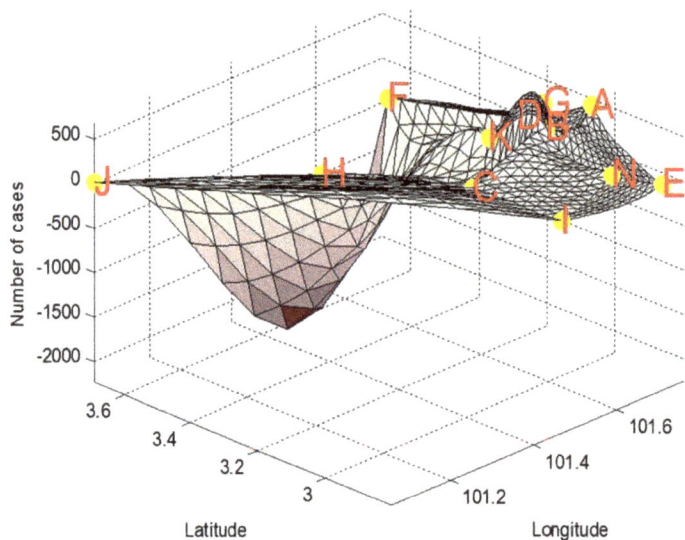

Figure 18. Interpolated surface without positivity preservation.

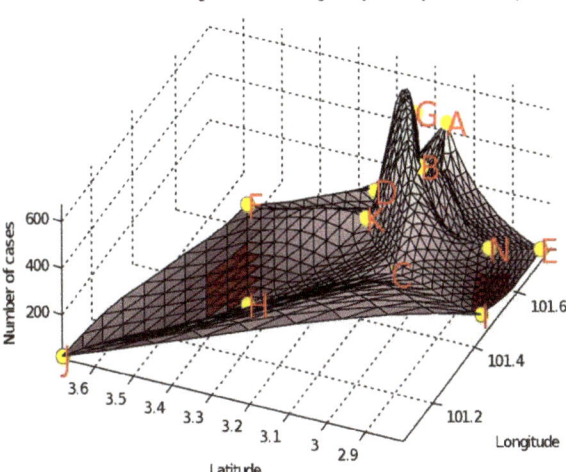

Figure 19. Interpolated surface with positivity preservation.

6. Conclusions

Zhu and Han [44] proposed new cubic Bernstein–Bézier basis functions defined on a triangular domain. We implemented quartic triangular bases (with ten control points) for scattered data interpolation. This new quartic basis makes it possible to avoid the optimization problem that appears when the quartic Bézier triangular is used for scattered data interpolation. From the results, we can see that the proposed scheme in this study outperformed the quartic Bézier triangular, having the smallest maximum error, higher R^2, and requiring only 12.5% of the CPU time needed by the quartic Bézier triangular scheme. This is very significant, especially when the goal is to reconstruct surfaces from large scattered data sets. Furthermore, based on a comparison against the Shepard triangular for scattered data, the proposed scheme was also superior to the schemes of Cavoretto et al. [6], Dell'Accio et al. [12,13] and Dell'Accio and Di Tommaso [11]. Finally, we constructed a positive interpolant based on the proposed quartic triangular spline to preserve the positivity of scattered data. Numerical results suggest that the proposed scheme is better than existing schemes, especially in terms of CPU time—our proposed scheme requires less computation time than positivity schemes proposed by Piah et al. [36] and Saaban et al. [35]. Finally, we implemented our proposed positivity-preserving interpolation to visualize COVID-19 cases in Selangor State and Klang Valley, Malaysia. The resulting surfaces were smooth and positive everywhere. Future works will focus on the construction of a quintic Zhu and Han spline for scattered data interpolation with quintic precision as well as shape-preserving interpolation (e.g., positivity-preserving and range-restricted interpolation). This can be achieved by extending the main idea from Karim et al. [46]. Another potential study could be a comparison between the use of a CPU and a graphical processing unit (GPU) for large scattered data sets. Finally, the proposed scheme can also be applied to visualize large sets of scattered data, such as from geophysical data, medical imaging, and total COVID-19 cases around the world.

Author Contributions: Conceptualization, V.T.N.; Data curation, S.A.A.K., A.S. and V.T.N.; Formal analysis, S.A.A.K., A.S. and V.T.N.; Funding acquisition, S.A.A.K.; Investigation, S.A.A.K.; Methodology, S.A.A.K. and A.S.; Resources, S.A.A.K. and V.T.N.; Software, S.A.A.K. and A.S.; Validation, S.A.A.K., A.S. and V.T.N.; Visualization, S.A.A.K., A.S. and V.T.N.; Writing—original draft, S.A.A.K.; Writing—review and editing, S.A.A.K., A.S. and V.T.N. All authors have read and agreed to the published version of the manuscript.

Funding: This research is fully supported by Universiti Teknologi PETRONAS (UTP) and Ministry of Education (MOE), Malaysia for the financial support received in the form of a research grant: FRGS/1/2018/STG06/UTP/03/1/015MA0-020 (New Rational Quartic Spline For Image Refinement) and YUTP: 0153AA-H24 (Spline Triangulation for Spatial Interpolation of Geophysical Data).

Conflicts of Interest: The authors declare no conflict of interest.

References

1. Ali, F.A.M.; Karim, S.A.A.; Bin Saaban, A.; Hasan, M.K.; Ghaffar, A.; Nisar, K.S.; Baleanu, D. Construction of Cubic Timmer Triangular Patches and its Application in Scattered Data Interpolation. *Mathematics* **2020**, *8*, 159. [CrossRef]
2. Bracco, C.; Gianelli, C.; Sestini, A. Adaptive scattered data fitting by extension of local approximations to hierachical splines. *Comput. Aided Geom. Des.* **2017**, *52–53*, 90–105. [CrossRef]
3. Borne, S.L.; Wende, M. Domain decomposition methods in scattered data interpolation with conditionally positive definite radial basis functions. *Comput. Math. Appl.* **2018**, *77*, 1178–1196. [CrossRef]
4. Bozzini, M.; Lenarduzzi, L.; Rossini, M. Polyharmonic splines: An approximation method for noisy scattered data of extra-large size. *Appl. Math. Comput.* **2010**, *216*, 317–331. [CrossRef]
5. Brodlie, K.W.; Asim, M.R.; Unsworth, K. Constrained Visualization Using the Shepard Interpolation Family. *Comput. Graph. Forum* **2005**, *24*, 809–820. [CrossRef]
6. Cavoretto, R.; De Rossi, A.; Dell'Accio, F.; Di Tommaso, F. Fast computation of triangular Shepard interpolants. *J. Comput. Appl. Math.* **2019**, *354*, 457–470. [CrossRef]
7. Chan, E.; Ong, B. Range restricted scattered data interpolation using convex combination of cubic Bézier triangles. *J. Comput. Appl. Math.* **2001**, *136*, 135–147. [CrossRef]
8. Chang, L.; Said, H. A C^2 triangular patch for the interpolation of functional scattered data. *Comput. Des.* **1997**, *29*, 407–412. [CrossRef]
9. Draman, N.N.C.; Karim, S.A.A.; Hashim, I. Scattered Data Interpolation Using Rational Quartic Triangular Patches With Three Parameters. *IEEE Access* **2020**, *8*, 44239–44262. [CrossRef]
10. Fasshauer, G.E. *Meshfree Approximation Methods with Matlab*; World Scientific Publishing Co Pte Ltd.: Singapore, 2007.
11. Dell'Accio, F.; Di Tommaso, F. On the hexagonal Shepard method. *Appl. Numer. Math.* **2020**, *150*, 51–64. [CrossRef]
12. Dell'Accio, F.; Di Tommaso, F.; Nouisser, O.; Zerroudi, B. Increasing the approximation order of the triangular Shepard method. *Appl. Numer. Math.* **2018**, *126*, 78–91. [CrossRef]
13. Dell'Accio, F.; Di Tommaso, F.; Hormann, K. On the approximation order of triangular Shepard interpolation. *IMA J. Numer. Anal.* **2015**, *36*, 359–379. [CrossRef]
14. Crivellaro, A.; Perotto, S.; Zonca, S. Reconstruction of 3D scattered data via radial basis functions by efficient and robust techniques. *Appl. Numer. Math.* **2017**, *113*, 93–108. [CrossRef]
15. Chen, Z.; Cao, F. Scattered data approximation by neural networks operators. *Neurocomputing* **2016**, *190*, 237–242. [CrossRef]
16. Zhou, T.; Li, Z. Scattered Noisy Data Fitting Using Bivariate Splines. *Procedia Eng.* **2011**, *15*, 1942–1946. [CrossRef]
17. Zhou, T.; Li, Z. Scattered noisy Hermite data fitting using an extension of the weighted least squares method. *Comput. Math. Appl.* **2013**, *65*, 1967–1977. [CrossRef]
18. Qian, J.; Wang, F.; Zhu, C. Scattered data interpolation based upon bivariate recursive polynomials. *J. Comput. Appl. Math.* **2018**, *329*, 223–243. [CrossRef]
19. Liu, Z. Local multilevel scattered data interpolation. *Eng. Anal. Bound. Elements* **2018**, *92*, 101–107. [CrossRef]
20. Joldes, G.; Chowdhury, H.A.; Wittek, A.; Doyle, B.J.; Miller, K. Modified moving least squares with polynomial bases for scattered data approximation. *Appl. Math. Comput.* **2015**, *266*, 893–902. [CrossRef]
21. Lai, M.-J.; Meile, C. Scattered data interpolation with nonnegative preservation using bivariate splines and its application. *Comput. Aided Geom. Des.* **2015**, *34*, 37–49. [CrossRef]

22. Schumaker, L.L.; Speleers, H. Nonnegativity preserving macro-element interpolation of scattered data. *Comput. Aided Geom. Des.* **2010**, *27*, 245–261. [CrossRef]
23. Karim, S.A.A.; Saaban, A.; Hasan, M.K.; Sulaiman, J.; Hashim, I. Interpolation using cubic Bézier triangular patches. *Int. J. Adv. Sci. Eng. Inf. Technol.* **2018**, *8*, 1746–1752. [CrossRef]
24. Karim, S.A.A.; Saaban, A.; Skala, V.; Ghaffar, A.; Nisar, K.S.; Baleanu, D. Construction of new cubic Bézier-like triangular patches with application in scattered data interpolation. Available online: https://link.springer.com/content/pdf/10.1186/s13662-020-02598-w.pdf (accessed on 29 March 2020).
25. Karim, S.A.B.A.; Saaban, A. Visualization Terrain Data Using Cubic Ball Triangular Patches. *MATEC Web Conf.* **2018**, *225*, 06023. [CrossRef]
26. Said, H.B.; Rahmat, R.W. A Cubic Ball Triangular Patch for Scattered Data Interpolation. *J. Phys. Sci.* **1995**, *5*, 89–101.
27. Feng, R.; Zhang, Y. Piecewise Bivariate Hermite Interpolations for Large Sets of Scattered Data. *J. Appl. Math.* **2013**, *2013*, 1–10. [CrossRef]
28. Sun, Q.; Bao, F.; Zhang, Y.; Duan, Q. A bivariate rational interpolation based on scattered data on parallel lines. *J. Vis. Commun. Image Represent.* **2013**, *24*, 75–80. [CrossRef]
29. Goodman, T.N.T.; Said, H.B. A C^1- triangular interpolation suitable for scattered data interpolation. *Commun. Appl. Numer. Methods* **1991**, *7*, 479–485. [CrossRef]
30. Foley, T.A.; Opitz, K. Hybrid cubic Bézier triangle patches. In *Mathematical Methods in Computer Aided Geometric Design II*; Lyche, T., Schumaker, L.L., Eds.; Academic Press: Cambridge, MA, USA, 1992; pp. 275–286.
31. Goodman, T.; Said, H. Shape preserving properties of the generalised ball basis. *Comput. Aided Geom. Des.* **1991**, *8*, 115–121. [CrossRef]
32. Goodman, T.; Said, H. Properties of generalized Ball curves and surfaces. *Comput. Aided Des.* **1991**, *23*, 554–560. [CrossRef]
33. Hussain, M.Z.; Hussain, M. C^1 positivity preserving scattered data interpolation using rational Bernstein-Bézier triangular patch. *J. Appl. Math. Comput.* **2009**, *35*, 281–293. [CrossRef]
34. Goodman, T.; Said, H.; Chang, L. Local derivative estimation for scattered data interpolation. *Appl. Math. Comput.* **1995**, *68*, 41–50. [CrossRef]
35. Saaban, A.; Piah, A.R.M.; Majid, A.A.; Chang, L.H.T. G^1 scattered data interpolation with minimized sum of squares of principle curvatures. In Proceedings of the International Conference on Computer Graphics, Imaging and Visualization (CGIV'05), Beijing, China, 26–29 July 2005; pp. 385–390.
36. Piah, A.R.M.; Saaban, A.; Majid, A.A. Range restricted positivity-preserving scattered data interpolation. *Malays. J. Fundam. Appl. Sci.* **2014**, *2*, 63–75. [CrossRef]
37. Hussain, M.; Majid, A.A.; Hussain, M.Z. Convexity-preserving Bernstein–Bézier quartic scheme. *Egypt. Inform. J.* **2014**, *15*, 89–95. [CrossRef]
38. Hussain, M.; Hussain, M.Z.; Buttar, M. C^1 Positive Bernstein-Bézier Rational Quartic Interpolation. *Int. J. Math. Models Methods Appl. Sci.* **2014**, *8*, 9–21.
39. Farin, G. *Curves and Surfaces for CAGD: A Practicle Guide*, 5th ed.; Palmer, C., Ed.; Morgan Kaufmann: San Diego, CA, USA, 2001.
40. Franke, R. Scattered Data Interpolation: Tests of Some Method. *Math. Comput.* **1982**, *38*, 181. [CrossRef]
41. Franke, R.; Nielson, G.M. Scattered data interpolation of large Sets of scattered data. *Int. J. Numer. Methods Eng.* **1980**, *15*, 1691–1704. [CrossRef]
42. Franke, R.; Nielson, G.M. Scattered Data Interpolation and Applications: A Tutorial and Survey; Scattered data interpolation and applications: A tutorial and survey. In *Geometric Modelling: Methods and Applications*; Hagen, H., Roller, D., Eds.; Springer Science and Business Media LLC: Berlin/Heidelberg, Germany, 1991; pp. 131–160.
43. Lodha, S.; Franke, R. Scattered Data Techniques for Surfaces. In Proceedings of the Scientific Visualization Conference (dagstuhl '97), Dagstuhl, Germany, 9–13 June 1997; pp. 189–230.
44. Zhu, Y.; Han, X. A class of $\alpha\beta\gamma$-Bernstein–Bézier basis functions over triangular domain. *Appl. Math. Comput.* **2013**, *220*, 446–454. [CrossRef]

45. Press statement by the Director-General of Health Malaysia, Ministry of Health Malaysia. Available online: https://kpkesihatan.com (accessed on 15 April 2020).
46. Karim, S.A.A.; Saaban, A.; Skala, V. Range-Restricted Surface Interpolation Using Rational Bi-Cubic Spline Functions with 12 Parameters. *IEEE Access* **2019**, *7*, 104992–105007. [CrossRef]

© 2020 by the authors. Licensee MDPI, Basel, Switzerland. This article is an open access article distributed under the terms and conditions of the Creative Commons Attribution (CC BY) license (http://creativecommons.org/licenses/by/4.0/).

Article

Utilization of the Brinkman Penalization to Represent Geometries in a High-Order Discontinuous Galerkin Scheme on Octree Meshes

Nikhil Anand *, Neda Ebrahimi Pour *, Harald Klimach * and Sabine Roller *

Simulation Techniques and Scientific Computing, Department Mechanical Engineering, University of Siegen, 57076 Siegen, Germany
* Correspondence: nikhil.anand@uni-siegen.de (N.A.); neda.epour@uni-siegen.de (N.E.P.); harald.klimach@uni-siegen.de (H.K.); sabine.roller@uni-siegen.de (S.R.)

Received: 1 July 2019; Accepted: 3 September 2019; Published: 5 September 2019

Abstract: We investigate the suitability of the Brinkman penalization method in the context of a high-order discontinuous Galerkin scheme to represent wall boundaries in compressible flow simulations. To evaluate the accuracy of the wall model in the numerical scheme, we use setups with symmetric reflections at the wall. High-order approximations are attractive as they require few degrees of freedom to represent smooth solutions. Low memory requirements are an essential property on modern computing systems with limited memory bandwidth and capability. The high-order discretization is especially useful to represent long traveling waves, due to their small dissipation and dispersion errors. An application where this is important is the direct simulation of aeroacoustic phenomena arising from the fluid motion around obstacles. A significant problem for high-order methods is the proper definition of wall boundary conditions. The description of surfaces needs to match the discretization scheme. One option to achieve a high-order boundary description is to deform elements at the boundary into curved elements. However, creating such curved elements is delicate and prone to numerical instabilities. Immersed boundaries offer an alternative that does not require a modification of the mesh. The Brinkman penalization is such a scheme that allows us to maintain cubical elements and thereby the utilization of efficient numerical algorithms exploiting symmetry properties of the multi-dimensional basis functions. We explain the Brinkman penalization method and its application in our open-source implementation of the discontinuous Galerkin scheme, Ateles. The core of this presentation is the investigation of various penalization parameters. While we investigate the fundamental properties with one-dimensional setups, a two-dimensional reflection of an acoustic pulse at a cylinder shows how the presented method can accurately represent curved walls and maintains the symmetry of the resulting wave patterns.

Keywords: high-order methods; Brinkman penalization; discontinuous Galerkin methods; embedded geometry; high-order boundary; IMEX Runge–Kutta methods

1. Introduction

In simulations of fluid motion for engineering scenarios, we generally need to deal with obstacles or containment of a non-trivial shape. In mesh-based schemes, we have two options to represent such geometries: we can try to align the mesh with the geometries, such that the walls build a boundary of the mesh or we try to embed the boundary conditions inside the mesh elements. The first option eases the formulation of boundary conditions and their treatment in the scheme [1]. The second option avoids the need to adapt the mesh to the, possibly complex, geometry [2]. Correctly aligning the mesh with arbitrary geometries in the first option can become cumbersome for high-order approximations. Thus, the embedding method is attractive for high-order schemes. Another application area, where the embedded boundaries provide a benefit, are moving geometries, as the need for new meshes can be avoided during simulations.

High-order discretization schemes can represent smooth solutions with few degrees of freedom. This is an essential property for algorithms on modern computing systems as the memory bandwidth is a strongly limiting factor on new systems, due to the widening memory gap. A numerical scheme that allows for high-order approximations of the solution is the discontinuous Galerkin finite element method. In this method, the solution within elements is represented by a function series (usually a polynomial series). In this work, we are concerned with a high-order discontinuous Galerkin scheme and the embedded geometry representation within it. Besides the possibility to use high-order approximations, the discontinuous Galerkin scheme also offers a relatively loose coupling between elements, resulting in a high computational locality, which in turn is advantageous for modern parallel computing systems. Discontinuous Galerkin methods are, therefore, increasingly popular and relevant.

Peskins [3] was one of the first scientists trying to impose immersed boundaries for his investigations. For his studies, he simulated the flow around heart valves considering the incompressible Navier–Stokes equations while introducing the immersed boundaries, using an elastic model and applying forces to the fluid, thus changing the momentum equations. His work was extended by Saiki and Biringen [4], and they considered feedback forces for the immersed boundaries to represent a rigid body while using an explicit time-stepping, hence resulting in stiff problems and very small time-stepping for the simulation. An important fact, which makes immersed boundary methods more attractive, is the introduction of the effect of the geometry in the governing equations themselves. Embedding the boundaries in the mesh relaxes the requirements on the elements, and using simple elements allows for efficient numerical algorithms that can, for example, exploit inherent symmetric properties of the discretization. The additionally introduced terms can either be considered in the numerical discretization or the continuous equations. Applying forcing terms in the discretization allows for better control of the numerical accuracy and the conservation properties of the used discretization method; on the other hand, the generality and flexibility of these methods disappear when considering different solvers using different discretization methods. In contrast, the volume penalization method imposes additional forcing penalty terms on the continuous equations, while the discretization is done as usual [5]. The Brinkman Penalization Method (BPM) is one of these methods. It was originally developed by Arquis and Caltagirone [6] for numerical simulations of isothermal obstacles in incompressible flows. The idea is to model the obstacle as a porous material, with material properties approaching zero. The major benefit of this method is error estimation, which can be rigorously predicted in terms of the penalization parameters [7]. Furthermore, the boundary conditions can be enforced to a precision, without changing the numerical discretization of the scheme. Kevlahan and Ghidaglia already applied this method for incompressible flows, while considering a non-moving, as well as a moving geometry. They used a pseudo-spectral method [8] in their works.

Liu and Vasilyev employed the volume penalization for the compressible Navier–Stokes equations. In their publication [9], they discussed a 1D and a 2D test case. They used a wavelet method for the discretization and showed error convergence and resulting pressure perturbations for acoustic setups. In other investigations, various numerical discretization methods were used, which showed promising results using the Brinkman penalization method. In [10,11], the pseudo-spectral methods, in [9], wavelet, and in [12], the finite volume/finite element methods were used. However, as far as we know, no work on this kind of penalization in the context of high-order discontinuous Galerkin methods for compressible Navier–Stokes equations has been done so far. Thus, this paper will look into the Brinkman penalization employed within a high-order discontinuous Galerkin solver. Our implementation is available in our open-source solver Ateles [13].

2. Numerical Method

The flow of compressible viscous fluids can mathematically be described by the Navier–Stokes equations governing the conservation of mass, momentum, and energy. In this section, we describe the compressible Navier–Stokes equation with the Brinkman penalization method to model solid obstacles as proposed in [9]. We apply this penalization in the frame of the Discontinuous Galerkin (DG) method and introduce this method also briefly in this section. The additional source terms introduced by the penalization increase the stiffness of the scheme considerably, and the last part of this section discusses how this can be overcome by an implicit-explicit time integration scheme.

2.1. The Compressible Navier–Stokes Equation

The Navier–Stokes equations describe the motion of fluids and model the conservation of mass, momentum, and energy. The non-dimensional compressible equations in conservative form can be written as:

$$\frac{\partial \rho}{\partial t} = -\nabla \cdot \mathbf{m}, \tag{1}$$

$$\frac{\partial m_i}{\partial t} + \sum_{j=1}^{3} \frac{\partial}{\partial x_j}(m_i v_j + p\delta_{ij}) - \frac{1}{Re}\sum_{j=1}^{3}\frac{\partial}{\partial x_j}\tau_{ij} = 0 \quad i = 1,2,3 \tag{2}$$

$$\frac{\partial \rho e}{\partial t} + \nabla \cdot \left[\left(e + \frac{p}{\rho}\right)\mathbf{m}\right] - \frac{1}{Re}\sum_{j=1}^{2}\frac{\partial}{\partial x_j}\left(\sum_{i=1}^{2}\tau_{ij}v_i - \frac{1}{\gamma-1}\frac{\mu}{Pr}T\right) = 0 \tag{3}$$

where the conserved quantities are the density ρ, the momentum $\mathbf{m} = \rho \mathbf{v}$, and the total energy density e, given by the sum of kinetic and internal energy density:

$$e = \frac{1}{2}|\mathbf{v}|^2 + \frac{p}{\rho(\gamma - 1)}. \tag{4}$$

where $\mathbf{v} = (v_1, v_2, v_3)^T$ is the velocity vector, δ_{ij} is the Kronecker delta, Re is the reference Reynolds number, and Pr the reference Prandtl number. γ stands for the isentropic expansion factor, given by the heat capacity ratio of the fluid, and T denotes the temperature. Viscous effects are described by the shear stress tensor:

$$\tau_{ij} = \mu \left(\frac{\partial v_i}{\partial x_j} + \frac{\partial v_j}{\partial x_i}\right) \tag{5}$$

and the dynamic viscosity μ.

To close the system, we use the ideal gas law as the equation of state, which yields the following relation:

$$p = \rho RT. \tag{6}$$

where R represents the gas constant.

2.2. The Brinkman Penalization

Penalization schemes employ additional, artificial terms to the equations in regions where the flow is to be inhibited (penalized). In the conservation of momentum and energy, we can make use of local source terms that penalize deviations from the desired state. With the Brinkman penalization, we also inhibit mass flow through obstacles by introducing the Brinkman porosity model and using a low porosity, where obstacles are to be found. Extending the compressible Navier–Stokes equations from Section 2.1 by the penalization terms, we obtain Equations (7) and (9).

$$\frac{\partial \rho}{\partial t} = -\nabla \cdot \left[1 + \left(\frac{1}{\phi} - 1\right)\chi\right]\mathbf{m}, \quad (7)$$

$$\frac{\partial m_i}{\partial t} + \sum_{j=1}^{3}\frac{\partial}{\partial x_j}\left(m_i v_j + p\delta_{ij}\right) - \frac{1}{Re}\sum_{j=1}^{2}\frac{\partial}{\partial x_j}\tau_{ij}$$
$$= -\frac{\chi}{\eta}(v_i - U_{oi}) \quad i = 1,2,3, \quad (8)$$

$$\frac{\partial \rho e}{\partial t} + \nabla \cdot \left[\left(e + \frac{p}{\rho}\right)\mathbf{m}\right] - \frac{1}{Re}\sum_{j=1}^{2}\frac{\partial}{\partial x_j}\left(\sum_{i=1}^{2}\tau_{ij}v_i - \frac{1}{\gamma-1}\frac{\mu}{Pr}T\right)$$
$$= -\frac{\chi}{\eta_T}(T - T_o). \quad (9)$$

The obstacle has the porosity ϕ, the velocity U_o, and the temperature T_o. The strength of the source terms can be adjusted by the viscous permeability η and the thermal permeability η_T. The masking function χ describes the geometry of obstacles and is zero outside of obstacles and one inside. It is also referred to as the characteristic function. It is capable of dealing not only with complex geometries but also with variations in time.

$$\chi(\mathbf{x},t) = \begin{cases} 1, & \text{if } x \in \text{obstacle.} \\ 0, & \text{otherwise.} \end{cases} \quad (10)$$

To represent a solid wall for compressible fluids properly, Liu et al. [9] stated that the porosity ϕ should be as small as possible, i.e., $0 < \phi << 1$. They scaled the permeabilities with the porosity and introduced according scaling factors α and α_T. The permeabilities were then defined by $\eta = \alpha\phi$ and $\eta_T = \alpha_T\phi$. With these relations, Liu et al. [9] found a modeling error of $O(\eta^{1/2}\phi)$ for resolved boundary layers in the material and $O((\eta/\eta_T)^{1/4}\phi^{3/4})$ for non-resolved boundary layers. In both cases, the error was dominated by the porosity. Nevertheless, the error can still be minimized with sufficiently small viscous permeabilities η.

Moreover, small values of the porosity caused stability issues and imposed a heavy time-step restriction with our numerical scheme. With the introduction of ϕ, the eigenvalues of the hyperbolic system changed, which has adverse effects on stability. The eigenvalues of the system of equations along with penalization terms [9] are given by the following characteristic equation:

$$-(\lambda - u)^3 + \left[c^2 + \frac{u^2}{2}(\phi^{-1} - 1)(\gamma - 3)\right](\lambda - u) - c^2 u(\phi^{-1} - 1)(\gamma - 1) = 0, \quad (11)$$

where $c = (\gamma p/\rho)^{1/2}$ and γ, p, ρ, and u are the ratio of specific heat, pressure, density, and velocity, respectively. For $\phi = 1$, the system of equations yields three eigenvalues $u, u + c, u - c$, which implies the speed of sound c in the medium, which is what we would like to achieve. However, with $0 < \phi << 1$, the eigenvalues can no longer be evaluated easily and are linked to ϕ, which causes problems for the hyperbolic part.

2.3. The Discontinuous Galerkin Discretization

In this section, we briefly introduce the semi-discrete form of the Discontinuous Galerkin finite element method (DG) for compressible inviscid flows. The compressible Euler equations were derived from the Navier–Stokes equations by neglecting diffusive terms. They still provide a model for the conservation of mass, momentum, and energy in the fluid and can be described in vectorial notation by:

$$\partial_t \mathbf{u} + \nabla \cdot \mathbf{F}(\mathbf{u}) = 0, \quad (12)$$

equipped with suitable initial and boundary conditions. Here, **u** is a vector of the conservative variables, and the flux function $\mathbf{F}(\mathbf{u}) = (\mathbf{f}(\mathbf{u}), \mathbf{g}(\mathbf{u}))^T$ for two spatial dimensions is given by:

$$\mathbf{u} = \begin{bmatrix} \rho \\ \rho u \\ \rho v \\ \rho E \end{bmatrix}, \mathbf{f}(\mathbf{u}) = \begin{bmatrix} \rho u \\ \rho u^2 + p \\ \rho u v \\ (\rho E + p)u \end{bmatrix}, \mathbf{g}(\mathbf{u}) = \begin{bmatrix} \rho v \\ \rho u v \\ \rho v^2 + p \\ (\rho E + p)v \end{bmatrix},$$

where $\rho, \mathbf{v} = (u,v)^T, E, p$ denotes the density, velocity vector, specific total energy, and pressure, respectively. The system is closed by the equation of state assuming the fluid obeys the ideal gas law with pressure defined as $p = (\gamma - 1)\rho \left(e - \frac{1}{2}(u^2 + v^2)\right)$; where $\gamma = \frac{c_p}{c_v}$ is the ratio of specific heat capacities and e is the total internal energy per unit mass.

The discontinuous Galerkin formulation of the above equation was obtained by multiplying it with a test function ψ and integrating it over the domain Ω. Thereafter, integration by parts was used to obtain the following weak formulation:

$$\int_\Omega \psi \frac{\partial \mathbf{u}}{\partial t} d\Omega + \oint_{\partial \Omega} \psi \mathbf{F}(\mathbf{u}) \cdot \mathbf{n} ds - \int_\Omega \nabla \psi \cdot \mathbf{F}(\mathbf{u}) d\Omega = 0, \forall \psi, \tag{13}$$

where ds denotes the surface integral. A discrete analogue of the above equation was obtained by considering a tessellation of the domain Ω into n closed, non-overlapping elements given by $T = \{\Omega_i | i = 1, 2, \ldots, n\}$, such that $\Omega = \cup_{i=1}^n \Omega_i$ and $\Omega_i \cap \Omega_j = \emptyset \forall i \neq j$. We define a finite element space consisting of discontinuous polynomial functions of degree $m \geq 0$ given by:

$$P^m = \{f \in [L^2(\Omega)]^m : f_{|\Omega_k} \in \mathbb{P}^m(\Omega_k) \forall \Omega_k \in \Omega\} \tag{14}$$

where $\mathbb{P}^m(\Omega_k)$ is the space of polynomials with largest degree m on element Ω_k. With the above definition, we can write the approximate solution $\mathbf{u}_h(\mathbf{x}, t)$ within each element using a polynomial of degree m:

$$\mathbf{u}_h(\mathbf{x}, t) = \sum_{i=1}^m \hat{u}_i \phi_i, \quad \psi_h(\mathbf{x}) = \sum_{i=1}^m \hat{v}_i \phi_i, \tag{15}$$

where the expansion coefficients $\hat{u}_i$ and $\hat{v}_i$ denote the degrees of freedom of the numerical solution and the test function, respectively. Notice that there is no global continuity requirement for $\mathbf{u}_h$ and ψ_h in the previous definition. Splitting the integrals in Equation (13) into a sum of integrals over elements Ω_i, we obtain the space-discrete variational formulation:

$$\sum_{i=1}^n \frac{\partial}{\partial t} \int_{\Omega_i} \psi_h \mathbf{u}_h d\Omega + \oint_{\partial \Omega_i} \psi_h \mathbf{F}(\mathbf{u}_h) \cdot \mathbf{n} ds - \int_{\Omega_i} \nabla \psi_h \cdot \mathbf{F}(\mathbf{u}_h) d\Omega = 0, \forall \psi_h \in P^m. \tag{16}$$

Due to the element local support of the numerical representation, the flux term is not uniquely defined at the element interfaces. The flux function is, therefore, replaced by a numerical flux function $\mathbf{F}^*(\mathbf{u}_h^-, \mathbf{u}_h^+, \mathbf{n})$, where $\mathbf{u}_h^-$ and $\mathbf{u}_h^+$ are the interior and exterior traces at the element face in the direction $\mathbf{n}$ normal to the interface. A choice of appropriate numerical flux can then be selected from several numerical flux schemes. For our simulations, we used the Lax–Friedrichs scheme for numerical flux.

For simplicity, we can re-write the equation above in matrix vector notation and obtain:

$$\frac{\partial}{\partial t} \hat{\mathbf{u}} = M^{-1} \left(S \cdot \mathbf{F}(\hat{\mathbf{u}}) - M^F \cdot \mathbf{F}(\hat{\mathbf{u}}) \right) =: rhs(\hat{\mathbf{u}}). \tag{17}$$

where M, S denote the mass and the stiffness matrices and M^F are the so-called face mass lumping matrices. The above obtained ordinary differential Equation (17) can be solved in time using any standard time-stepping method, e.g., a Runge–Kutta method.

In our implementation, we exploited the fact that we only used cubical elements. This choice of simple elements allowed for a tensor-product notation in the multi-dimensional basis functions. The symmetry of the elements enabled efficient dimension-by-dimension algorithms in the computation.

2.4. The Implicit-Explicit Time Discretization

As the penalization introduces stiff terms to the equations and for accuracy, we would want them to be as stiff as possible, we introduce an implicit time integration for those terms. With an otherwise explicit time integration scheme, this results in an implicit-explicit time-stepping scheme that can be achieved by splitting the right-hand side of the equations into an explicitly integrated part and an implicitly integrated part. Therefore, to perform time integration of the system, we use a Diagonally Implicit Runge–Kutta (DIRK) scheme with three explicit and four implicit stages as presented in [14]. The following section first considers a single implicit Euler step, to discuss the arising equations that need to be solved in each implicit stage of the higher order time discretization.

We denote the right-hand side by Q and employ the superscript ι for the implicit part and the superscript ξ for the explicit part. By using the conservative quantities as subscripts (ρ, m, and e), we can distinguish the right-hand sides for the different equations. Thus, we get:

$$\frac{\partial \rho}{\partial t} = Q_\rho^\xi + Q_\rho^\iota \tag{18}$$

$$\frac{\partial m_i}{\partial t} = Q_{m_i}^\xi + Q_{m_i}^\iota \tag{19}$$

$$\frac{\partial e}{\partial t} = Q_e^\xi + Q_e^\iota \tag{20}$$

and we chose the implicit parts as:

$$Q_\rho^\iota = 0 \tag{21}$$

$$Q_{m_i}^\iota = -\frac{\chi}{\eta}(u_i - U_{oi}) \tag{22}$$

$$Q_e^\iota = -\frac{\chi}{\eta_T}(T - T_o) \tag{23}$$

out of Equations (7) and (9).

This choice restricts the implicit computation to the local source terms, which can be evaluated pointwise. Unfortunately, the introduced Brinkman porosity in (7) affects the flux and introduces spatial dependencies. To avoid the need for the solution of an equation system across the whole domain for this dependency, the porosity part will be computed in the explicit time-stepping scheme.

Observation for the Implicit Part

Considering Equations (18)–(20) only with their implicit parts, we get the following equation system:

$$\frac{\partial \rho}{\partial t} = 0 \tag{24}$$

$$\frac{\partial m_i}{\partial t} = -\frac{\chi}{\eta}(u_i - U_{oi}) \tag{25}$$

$$\frac{\partial e}{\partial t} = -\frac{\chi}{\eta_T}(T - T_o). \tag{26}$$

Notice that these equations can be solved pointwise as no spatial derivatives appear.

A discretization of these equations in time with a Euler backward scheme yields the solvable equation system:

$$\frac{\rho(t+\Delta t) - \rho(t)}{\Delta t} = 0 \tag{27}$$

$$\frac{m_i(t+\Delta t) - m_i(t)}{\Delta t} = -\frac{\chi}{\eta}\left(u_i(t+\Delta t) - U_{oi}\right) \tag{28}$$

$$\frac{e(t+\Delta t) - e(t)}{\Delta t} = -\frac{\chi}{\eta_T}\left(T(t+\Delta t) - T_o\right). \tag{29}$$

Equation (27) trivially yields $\rho(t+\Delta t) = \rho(t)$. With the implied constant density, we can now solve the equation for the change in momentum (28) and arrive at an explicit expression for the velocity $u_i(t+\Delta t)$ at the next point in time:

$$\frac{\rho(t+\Delta t)u_i(t+\Delta t) - \rho(t)u_i(t)}{\Delta t} = -\frac{\chi}{\eta}\left(u_i(t+\Delta t) - U_{oi}\right) \tag{30}$$

$$u_i(t+\Delta t) = \frac{\rho(t)u_i(t) + \frac{\chi\Delta t}{\eta}U_{oi}}{\rho(t) + \frac{\chi\Delta t}{\eta}}. \tag{31}$$

Finally, density and velocity at the new point in time can be used to find the new temperature as well by substituting the above results in Equation (29) and solving for the temperature at the next point in time. We find:

$$T(t+\Delta t) = \frac{\frac{\chi\Delta t}{\eta_T}T_o + c_v\rho(t)T(t) + \frac{\rho(t)}{2}(u_i^2(t) - u_i^2(t+\Delta t))}{c_v\rho(t) + \frac{\chi\Delta t}{\eta_T}}. \tag{32}$$

where $u_i(t+\Delta t)$ is given by Equation (31).

Thus, this specific choice of terms for the implicit part of the time integration scheme yields a system that can be solved explicitly and without much additional computational effort. However, the implicit discretization allows for arbitrarily small values of η and η_T. A similar approach was developed by Jens Zudrop to model perfectly electrical conducting boundaries in the Maxwell equations, and more details can also be found in his thesis [15].

To solve the complete system, we then employed the diagonally implicit Runge–Kutta scheme with three explicit stages and four implicit stages [14]. It provides a scheme that is third order in time and L-stable.

Note, that while this approach overcomes time step limitations with respect to the permeabilities η and η_T, the porosity term changes the eigenvalues of the hyperbolic system and affects the stability.

3. Results and Discussion

To investigate the penalization scheme in our discontinuous Galerkin implementation, we first analyzed the fundamental behavior in two one-dimensional setups and then considered the scattering at a cylinder in a two-dimensional setup.

As explained in Section 2.2, the modeling error by the penalization for the compressible Navier–Stokes equations as found by Liu and Vasilyev [9] was expected to scale with the porosity ϕ by an exponent between 3/4 and one and with the viscous permeability η by an exponent between 1/4 and 1/2. To achieve low errors, you may, therefore, be inclined to minimize ϕ. However, with the implicit mixed explicit time integration scheme presented in Section 2.4, we can eliminate the stiffness issues due to small permeabilities with little additional costs, while the stability limitation by the porosity persists. Because of this, we deem it more feasible to utilize a small viscous permeability instead of a small porosity. At the same time, the relation between viscous permeability η and thermal permeability η_T gets small without overly large η_T. Therefore, we used a slightly different scaling than proposed by Liu and Vasilyev [9]. We introduce the scaling parameter β and define the permeabilities accordingly in relation to the porosity as follows.

$$\eta = \beta^2 \cdot \phi^2 \qquad (33)$$

$$\eta_T = 0.4\beta \cdot \phi \qquad (34)$$

Note, that we then expect the modeling error to be of size $O(\beta^{1/4}\phi^{3/4})$.

3.1. One-Dimensional Acoustic Wave Reflection

To assess how well the penalization scheme can capture the reflective nature of a solid wall, we used the reflection of an acoustic wave at the material. The initial pressure distribution is shown in Figure 1. It is described by the Gaussian pulse given in Equation (35) around its center at $x = 0.25$ in the left half of the domain ($x \leq 0.5$).

$$\rho' = u' = p' = \epsilon \exp\left[-\ln(2)\frac{(x-0.25)^2}{0.004}\right] \qquad (35)$$

For the amplitude ϵ of the wave, we used a value of $\epsilon = 10^{-3}$. The perturbations in density ρ', velocity u', and pressure p' from (35) are applied to a constant, non-dimensionalized state with a speed of sound of one. This results in the initial condition for the conservative variables density ρ, momentum m, and total energy e as described in: (36).

$$\rho = 1 + \rho', m = \rho u', e = \frac{1}{\gamma(\gamma-1)} + \frac{p'}{\gamma-1} + \frac{1}{2}\rho(u')^2 \qquad (36)$$

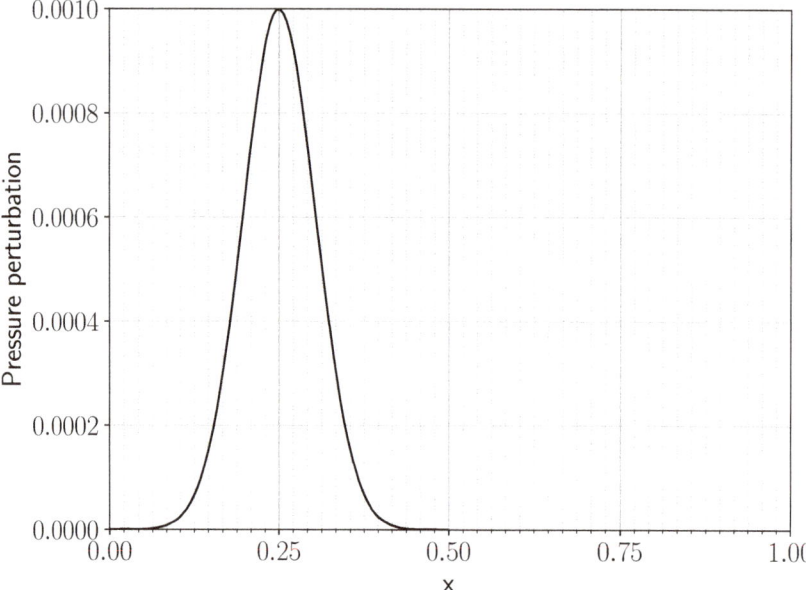

Figure 1. One-dimensional acoustic wave setup: the center of the initial pressure pulse is located at $x = 0.25$ and has an amplitude of $\epsilon = 10^{-3}$. Discretization by 48 elements as denoted by grid lines, and the right half of the domain ($x > 0.5$) is penalized. Note that the wall coincides with an element interface.

The penalization with porous medium is applied in the right half of the domain ($x > 0.5$). In acoustic theory, the reflection should be perfectly symmetric, and the reflected pulse should have the same shape and size, only with opposite velocity. This simple setup allows us to analyze the dampening of the reflected wave and induced phase errors. Reflected waves for different settings of β as defined in Equation (33) are shown in Figure 2. The pressure distribution for the reflection is shown for the state after a simulation time of 0.5. With linear acoustic wave transport and a speed of sound of one, the pulse should return to its original starting point, just with an opposite traveling direction. This symmetry makes it easy to judge both the loss in wave amplitude and the phase shift of the reflected pulse.

While the analytical result for a linear wave transport provided a good reference in general for the acoustic wave, it sufficiently deviated from the nonlinear behavior to limit its suitability for convergence analysis to small error values. Therefore, we compared the simulations with the penalization method to numerical results with traditional wall boundary conditions and a high resolution. This reference was computed with the same element length, but the domain ended at $x = 0.5$ with a wall boundary condition, and a maximal polynomial degree of 255 was used (256 degrees of freedom per element) to approximate the smooth solution. The resulting pressure profiles for different settings of β and a fixed porosity of $\phi = 1.0$ are shown in Figure 2. This illustrates how well the wave was reflected for different settings of β and that the solid wall reflection was well approximated for sufficiently small values of β. These numerical results were obtained with 48 elements and a maximal polynomial degree of 31 (32 degrees of freedom per element). Note that this setup aligns the wall interface with an element interface, where the discontinuity in the penalization is actually allowed by the numerical scheme. Later, we will discuss the changes observed, when moving the wall surface into the elements.

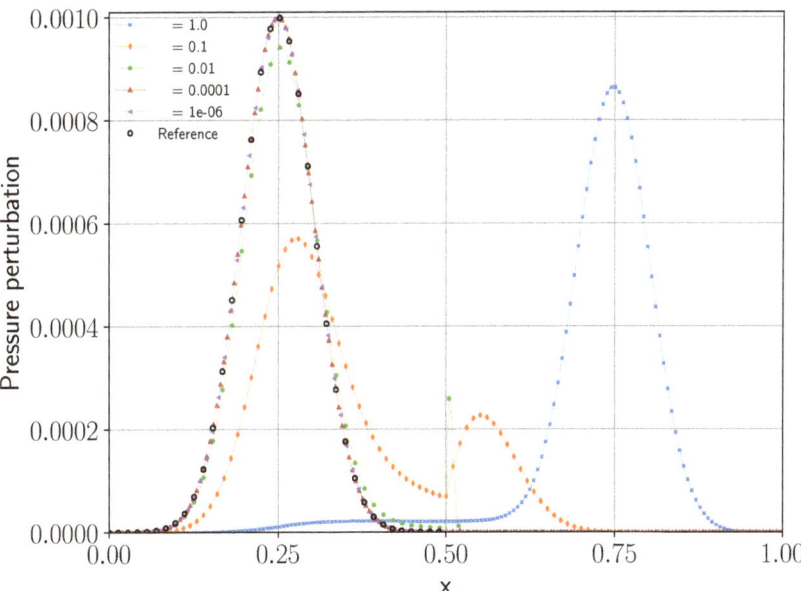

Figure 2. Plot for the pressure profile of the reflected wave at $t = 0.5$ for different scaling factors β. The numerical reference is obtained with a traditional wall boundary condition and a high resolution.

Figure 3 illustrates the impact of the porosity on the error in amplitude of the reflected wave for the same discretization with 48 elements and a maximal polynomial degree of 31. We plotted the error $e = (\epsilon - p'(t = 0.5))/\epsilon$ over porosity ϕ for various scaling parameters of β between one and 10^{-6}.

A scaling parameter of $\beta = 1$ means that the error is only driven by the porosity ϕ, and for large values of $\beta \geq 10^{-2}$, we observed the expected reduction in the error with decreasing porosity. However, with $\beta = 10^{-3}$, this comes eventually to an end (no improvements for $\phi < 2 \times 10^{-2}$), and for smaller values of β, no improvements for the error can be achieved by lowering the porosity anymore. As can be seen in this figure, a sufficiently small permeability can yield the accuracy as a small porosity.

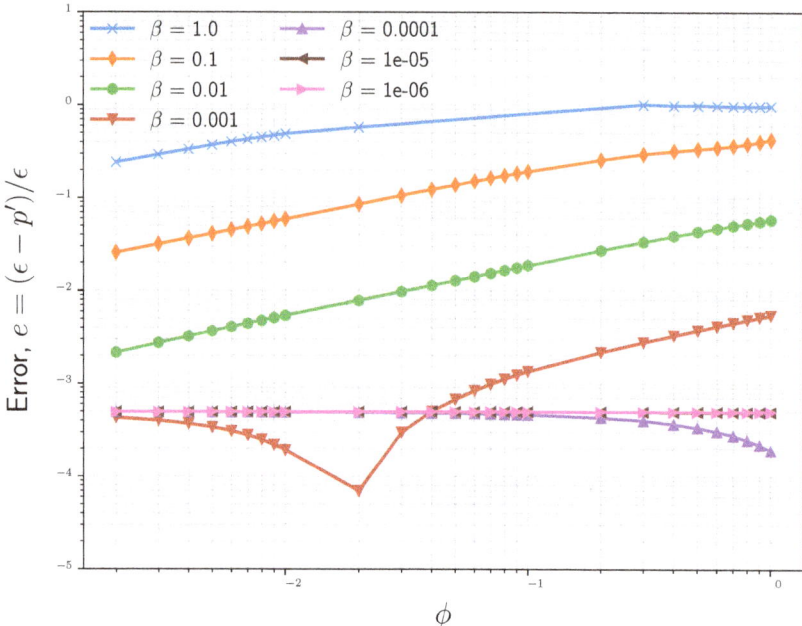

Figure 3. Plot of the error in the wave amplitude at $t = 0.5$ with decreasing porosity and different scaling factors β. The error e is given by the relative error in the pulse amplitude after the reflection at the wall.

In Figure 3 as well, for $\beta = 1e-3$, we observed a drop in error, and then it increased again with smaller ϕ to reach a convergence point finally. We would like to point out that this was expected to come from our numerical scheme using polynomials to represent the solution. Each data-point in the plot with varying ϕ and β represents a slightly different test case in terms of boundary layer thickness, as pointed out in Section 2.2. A sweet spot is reached when the degree of polynomial used for the simulation correctly captures the boundary layer in the problem. However, as we move further left from here, this sweet spot is slowly gone with further thinning of the boundary layer. With the same polynomial degree used, one would also expect to see this behavior for lines representing $\beta < 1e-3$ and correspondingly larger ϕ. This is exactly what we also see for $\beta = 1e-4$ and the value of ϕ close to 1.0. For all other lines in the plot, this spot does not fall within the range of the figure.

By using the implicit mixed explicit scheme from Section 2.4, it is possible to exploit drastically smaller values for the permeabilities to cover up the lack of porosity in the penalization. On the other hand, the porosity cannot that easily be treated in our discretization, and even moderate values of ϕ can have a dramatic impact on the time step restriction, due to the changed eigenvalues in the hyperbolic part of the equations.

Next, we performed a convergence analysis shifting the position of the wall such that it intersected the element at different locations. The reason for performing such an analysis becomes imperative when we consider the high-order numerical scheme used. We represented the solution state within an element using polynomials. For the pointwise evaluation of the nonlinear terms, we employed the Gaussian integration points, at which also the masking function of the penalization needs to be evaluated. Within an element, these integration points were scattered, being more concentrated on the element interface, and were rather sparse at the center. Therefore, in a peculiar validation test case like this one, when the wall was aligned with the element interface, due to the abundance of interpolation points, it had the advantage of being very precisely represented even for comparatively fewer degrees of freedom. In actual simulations, the wall interface may intersect an element at any point. We, therefore, also need to consider such intersections through the element and ensure the solution converges to the reference solution. The penalization method itself was not restrictive to any such limitations and could perfectly represent wall irrespective of its location within an element.

Thus, we performed and compared convergence analysis on two different discretizations, one where the wall lied at the element interface and a second where the wall intersected one element exactly in the middle. We would like to point out that the later scenario yielded a worst case estimate for the approximation of the jump in the masking function within an element. As explained, this simply came from scarce integration points lying around the center of element. For the following convergence analysis, we ignored the porosity (i.e., set $\phi = 1$) and used small permeabilities by choosing $\beta = 10^{-6}$. We also considered the L^2 error norm now in the fluid domain. As a reference solution, we employed a numerical simulation with a traditional wall boundary and a high maximal polynomial degree of 255 (256 degrees of freedom per element). The error was measured at $t = 0.5$ after the reflected wave reached its initial position again.

Figure 4 shows the L^2 norm of the error for the reflected pressure wave with a maximal polynomial degree of seven over an increasing number of elements (h-refinement). This plot compares the two discretizations explained. As expected, in Figure 4, we observe superior convergence behavior for the case when the wall lies at the element interface in comparison to the other case where the wall is crossing through the element center. However, for both cases, we observed a proper convergence towards the solution with a traditional solid wall boundary condition. The order of error convergence did not match the high-order discretization in either case, but this was expected due to the discontinuity introduced by the masking function of the penalization.

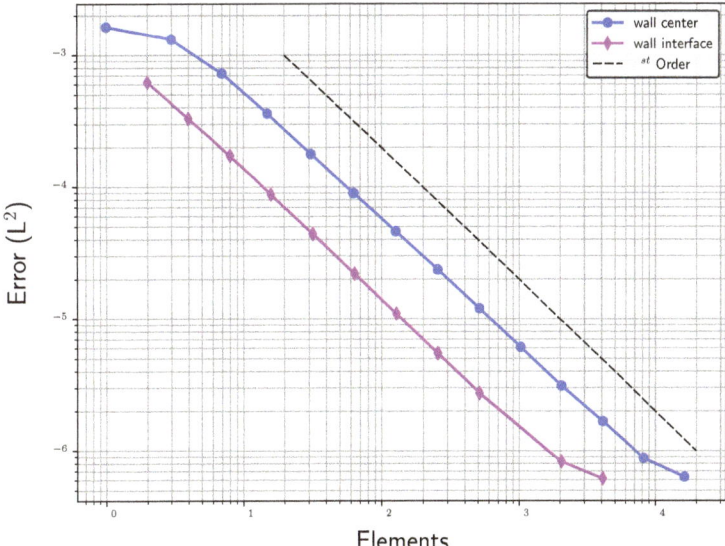

Figure 4. L^2-error for a polynomial degree of seven over an increasing number of elements (h-refinement).

Next, we performed another convergence study using the same two discretizations, but this time keeping the number of elements constant and increasing the order of polynomial representation within those elements.

Figure 5 shows the error convergence over the maximal polynomial degree in the discretization scheme (p-refinement) with the number of elements fixed to 24. Here, also, one observes a solution in both cases converging to the reference solution. While no spectral convergence was achieved for the discontinuous problem, a quadratic convergence can be observed. This shows that a high-order approximation was beneficial even with the discontinuous masking function for the penalization.

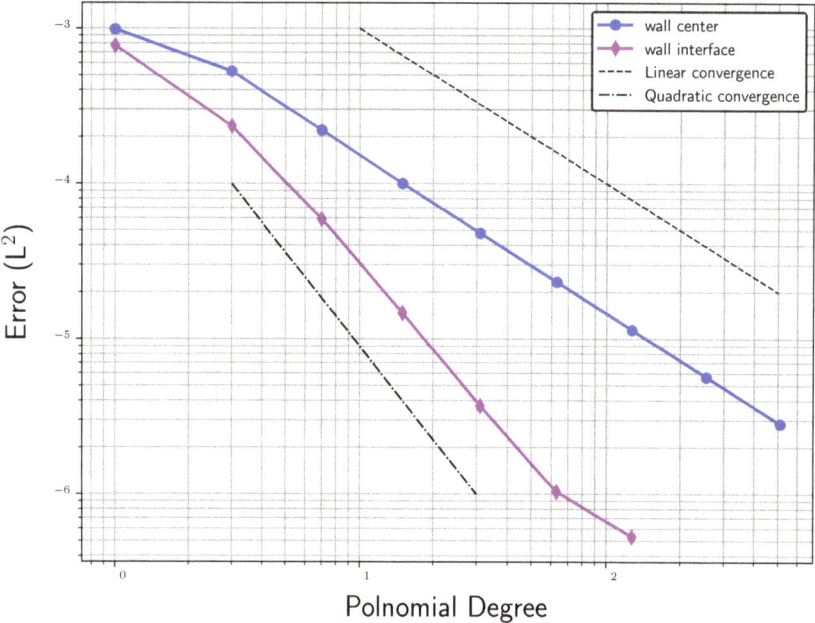

Figure 5. L^2-error for 24 elements over increasing maximal polynomial degree (p-refinement).

Finally, we also looked at the case where the wall was close to the element interface, but not exactly on it. This is a potential critical configuration for the numerical scheme being used, as the discontinuity close to the surface needs to be properly captured. We put the wall at 5% of the element length away from the element surface and measured the error as before, resulting in the graph shown in Figure 6. For this case as well, we observed a similar convergence rate as before, though the error was a little bit worse than with the wall on the interface.

For a smooth solution, the advantage of high-order methods to attain a numerical solution of a given quality using fewer degrees of freedom is well documented [16]. However, for a complex nonlinear problem with a discontinuity introduced by the porous medium, it is not so clear whether there is still a computational advantage by a high-order discretization. To investigate this, we ran the wall reflection problem for several orders and plotted the convergence with respect to the required computational effort, as seen in Figure 7. This test was performed starting with 16 elements in each data series, providing the leftmost point for the respective spatial scheme order. For subsequent data points, the number of elements was always increased by a factor of two up to the point where an error of 10^{-6} was achieved. Figure 7a depicts the observed L^2 error over the total number of degrees of freedom in the simulation. Here, we can see that for attaining a certain level of accuracy, the number of degrees of freedom required was always less when using a higher spatial order, even though the convergence rate did not increase with the scheme order. The high-order discretization, thus, allowed for memory-efficient computations, also in this case with a discontinuity present at the wall.

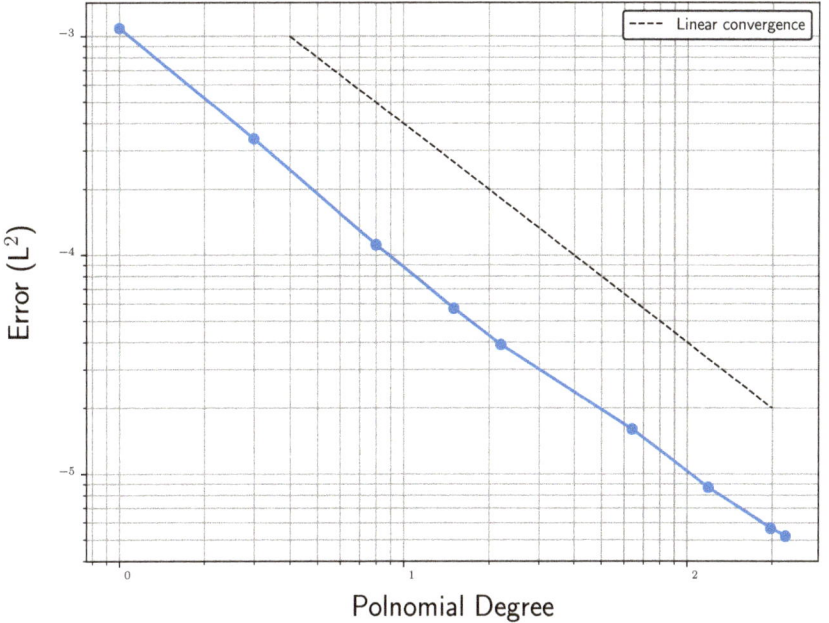

Figure 6. L^2-error for varying the polynomial degree. With a wall just 5% of the element length away from the element surface.

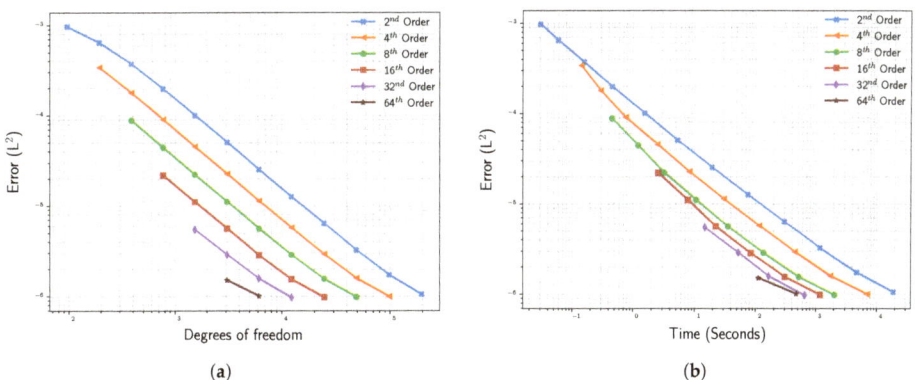

Figure 7. Behavior of the error in the reflected acoustic pulse with respect to computational effort. The figure on the left (**a**) shows the error convergence for various spatial orders over the required memory in terms of degrees of freedom. The right figure (**b**) shows the same runs, but now over the computational effort in terms of running time in seconds. All simulations were performed on a single node with 12 cores using 12 processes.

While for the memory consumption, there seems to be a clear benefit in high-order discretizations, it is not so clear whether this still holds for the required computing time. The time step limitation of the explicit scheme required more time steps for higher spatial scheme orders, increasing the computational effort to reach the desired simulation time. Additionally, the number of operations increased with higher orders due to the nonlinearity of the equations. Figure 7b shows the measured running times on a single computing node with 12 Intel Sandy-Bridge cores for the same runs. Again, the achieved accuracy is plotted, but this time over the observed running time in seconds. As can be seen, the advantage in terms of running times was not as clear as in terms of memory. However, we still observed faster times to the solution with higher spatial scheme orders, despite the increased number of time steps. In conclusion, we found some computational benefit from higher spatial scheme orders even in the presence of a discontinuity in this setup.

3.2. One-Dimensional Shock Reflection

After considering the reflection of an essentially linear acoustic wave, we now look into the reflection of a shock, where nonlinear terms play an important role. However, we neglected viscosity in this setup and only solved the inviscid Euler equations. The reflection of a one-dimensional shock wave at a wall was described and numerically investigated by Piquet et al. [17], for example. We used their setup to validate the penalization method in our discontinuous Galerkin setup, even though a high-order scheme is not ideal for the representation of shocks.

The downstream state in front of the shock (denoted by 1) is given in Table 1. The upstream state after the shock (denoted by 2) is then given by the Rankine–Hugoniot conditions for the shock Mach number Ma_s. These yield:

$$\frac{\rho_2}{\rho_1} = \frac{\gamma+1}{\gamma-1+2Ma_s^{-2}} \tag{37}$$

for the relation of densities ρ in up- and downstream of the shock and

$$\frac{p_2}{p_1} = \frac{2\gamma Ma_s^2 - (\gamma-1)}{\gamma+1} \tag{38}$$

for the relation of pressures p.

With these relations, the ratio of the upstream (p_3) and downstream (p_2) pressure for the reflected shock wave is [18]:

$$\frac{p_3}{p_2} = \frac{Ma_s^2(3\gamma-1) - 2(\gamma-1)}{2 + Ma_s^2(\gamma-1)} \tag{39}$$

For the computation of the velocity u_{rs} of the reflected shock wave, we considered Equation (40) [19].

$$u_{rs} = \frac{1}{Ma_s}\left(1 + \frac{2(Ma_s^2 - 1)}{(\gamma+1)/(\gamma-1)}\right)c_1 \tag{40}$$

Table 1. Shock state description.

Downstream speed of sound	c_1	1.0
Shock Mach number	Ma_s	1.2
Shock velocity	u_s	1.2
Downstream density	ρ_1	1.0
Downstream pressure	p_1	γ^{-1}
Downstream velocity	u_1	0.0
Isentropic coefficient	γ	1.4

With an incident shock wave velocity of $Ma_s = 1.2$, we obtained a pressure relation across the shock of 1.47826 from Equation (39). The shock was simulated in the unit interval $x \in [0,1]$ with the wall located at $x = 0.5$. Thus, half of the domain ($x \in [0.5, 1]$) was covered by the porous material to model the solid wall. The shock was initially located at $x = 0.25$.

For the numerical discretization, we used $256, 512, 1024$, and 2048 elements (n) in total ($\Delta x = 1/n$) and a scheme Order (O) of $32, 16, 8$, and 4, respectively. As explained in the inspection of the linear wave transport, our numerical scheme preferred strong permeabilities over penalization with the porosity. Therefore, we ignored porosity and chose $\phi = 1$, while the scaling factor from Equations (33) and (34) was set to a small value of $\beta = 10^{-6}$.

In Figure 8, the shock wave after its reflection is shown for different spatial resolutions. The discretizations with O(8) and 1024 elements and with O(4) and 2048 elements were chosen to have the same number of degrees of freedom, while the third discretization with O(8) and 2000 elements provided a high-resolution comparison.

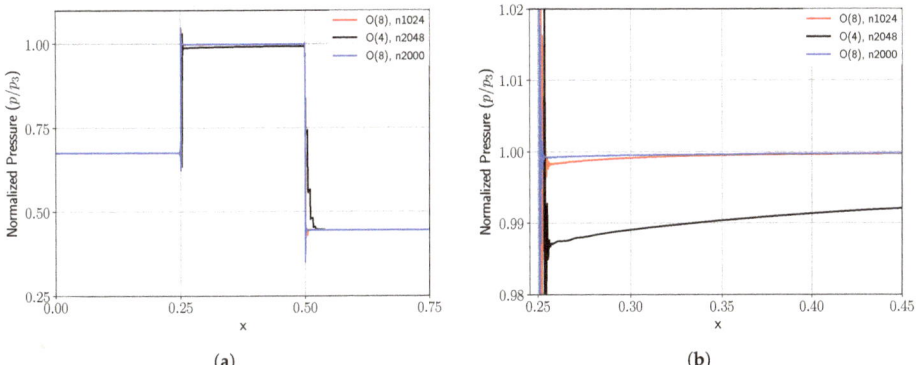

Figure 8. Different curves represent different discretizations using different scheme orders and a different number of elements. (**a**) Normalized pressure of the reflected shock wave. (**b**) Zoom of the reflected shock.

The exact solution for the normalized pressure (p^3/p^2) according to Equation (39) was 1.47826087. In Table 2, the ratio of the pressure (p^3/p^2), the relative error between the numerical, and the exact solution (error in p^3/p^2 in %) close to the shock, as well as the difference between the location of the shock wave after the reflection and its origin (Δx: phase shift) are listed. The table illustrates that with higher scheme order, but constant number of degrees of freedom, the error in the pressure ratio, as well as in the phase shift reduces considerably even for this discontinuous solution. From the obtained results, we can conclude that we achieved the same error as in [17], when using O(16) and 512 elements. As can be seen in Figure 8b, the plateau after the shock was not fully flat, but rather had a slope that asymptotically got close to the expected constant value. Except for the fourth-order approximation, this constant plateau was well obtained, but it remained slightly off the exact solution. This remaining error was also stated in the table as min_{error} and had a value of 0.0129% for $\beta = 10^{-6}$.

Table 2. Comparison of simulation results with the exact solution.

Test case	p^3/p^2	Error in p^3/p^2 in [%]	$\Delta x \cdot 10^{-4}$
n2048, O(4)	1.46053873	1.19885086	32.0161
n1024, O(8)	1.47642541	0.12416375	13.0319
n512, O(16)	1.47700446	0.08499256	8.1828
n256, O(32)	1.47714175	0.07570497	7.6228
n128, O(64)	1.47721414	0.07080803	6.4032
n2000, O(8)	1.47740998	0.05755990	7.0317
min_{error}	1.47806952	0.012944346	--

In Table 3, the results for the reflected shock wave are presented for the case that the wall is inside an element, instead of at its edge. Again, the error was reduced by an increased scheme order and a fixed total number of degrees of freedom. Notably, the error in the pressure ratio was reduced for small element counts in relation to the case where the wall coincided with an element interface. This was due to the fact that there was an additional element introduced here, and the element length was accordingly smaller. However, we can see that the phase shift of the shock was larger in this case. This can be attributed to the larger distance of the Gaussian integration points, which were used to represent the wall interface.

Table 3. Comparison of simulation results, when the porous material was located in the middle of the element with the exact solution.

Test Case	p^3/p^2	Error in p^3/p^2 in [%]	$\Delta x \cdot 10^{-4}$
n2049, O(4)	1.44333420	2.36268639	25.6373
n1025, O(8)	1.47333865	0.33297335	21.5178
n513, O(16)	1.47750687	0.05100577	15.4564
n257, O(32)	1.47751832	0.05023153	13.0052
min_{error}	1.47801754	0.01646083	--

For better comparison of the simulation results, Figure 9 illustrates the different test cases. The plot presents the solution, from the previous investigation, when using a scheme order of O(16). As a reference, we considered a no-slip wall, which was located at the same place as the porous material, while considering the same scheme order.

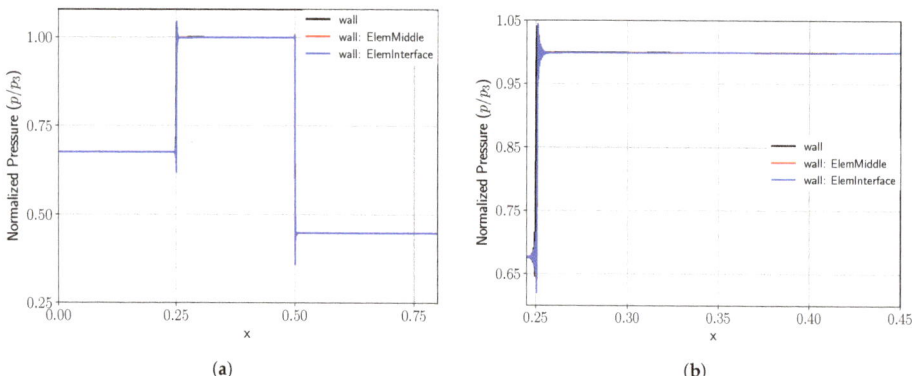

(a) (b)

Figure 9. Different curves represent different locations of the porous material in the element and the solution when using a no-slip wall. (a) illustrates the normalized pressure of the reflected shock wave, and (b) depicts a zoom-in of the front area of the reflected shock.

Figure 9 accentuates the close solution, when modeling the wall as a porous material, to the reference wall. As can be seen in Figure 9a, the high-order discretization introduced Gibbs oscillations around the shock, but otherwise, the discontinuity was well preserved by the numerical scheme. Some of those oscillations remained inside the modeled material (see Figure 9b) for the solid wall, but again, the discontinuity was well preserved. Further, the reflected shock exhibited an over- and under-shoot. Since the material was represented in polynomial space and according to the Gibbs phenomenon, we were limited to 9% deviation for physical correctness of the solution, we also computed those over- and under-shoots. For the material located inside an element, the overshoot was around 2.052% and the undershoot around 8.639%. Locating the porous material at the element interface resulted in 1.821% and 7.681%, respectively.

3.3. Scattering at a Cylinder

While the one-dimensional setups served well to demonstrate the basic numerical properties of the penalization scheme, they did not show the benefit of this approach. Only with multiple dimensions, the mesh generation was problematic for high-order schemes. Thus, we now turn to the scattering of a two-dimensional acoustic wave at a cylinder. The result was compared against the analytical solution of the linearized equations presented in [20]. In this case, the surface of the object was curved, and the wave did not only impinge in the normal direction of the obstacle. The expected symmetric scattering pattern of the reflected pulse eased the identification of numerical issues introduced by the modeling of the cylinder wall. Thus, this setting illustrated the treatment of curved boundaries in the high-order approximation scheme by penalization within simple square elements. The problem setup is depicted in Figure 10 and consisted of a cylinder of diameter $d = 1.0$ with its center lying at the point P with the coordinates $(10, 10)$.

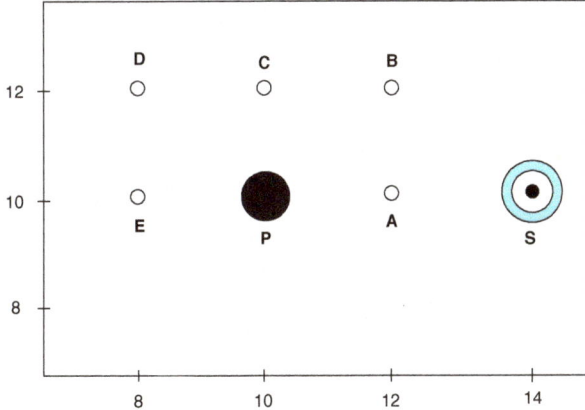

Figure 10. Test case setup for the wave scattering; only the section containing the cylindrical obstacle, the probing points, and the initial pulse is shown, and the actual computational domain is larger. The cylindrical obstacle is represented by the black circle located at $P(10, 10)$. Five observation points $(A, \ldots, E)$ around the obstacle are shown as circles. The initial pulse in pressure is indicated by the black dot with a turquoise circle around it located at $S(14, 10)$.

The initial condition prescribed a circular Gaussian pulse in pressure with its center at the point $S = (14, 10)$ and a half-width of 0.2. Thus, the initial condition for the perturbation of pressure is given by:

$$p' = \epsilon \exp\left[-\ln(2)\frac{(x-14)^2 + (y-10)^2}{0.04}\right]. \tag{41}$$

The amplitude $\epsilon = 10^{-3}$ of the pulse was chosen to be sufficiently small to nearly match the full compressible Navier–Stokes simulation with the linear reference solution.

The initial condition in terms of the conservative variables was given as:

$$\rho = \rho_0 + p', m_1 = m_2 = 0, e = c_p T - \frac{p}{\rho}. \tag{42}$$

Here, ρ_0 is the background density chosen as $\rho_0 = 1.0$. m_1, m_2 are the momentum in the x and y direction, respectively, and e is the total energy. T and c_p are the temperature and specific heat at constant pressure, respectively. The ratio of specific heats was chosen to be $\gamma = 1.4$. The Reynolds number used was $Re = 5 \times 10^5$, calculated using the diameter of the cylinder $d = 1.0$ as the characteristic length. Figure 10 shows the test case setup magnified around the area of interest.

The overall simulation domain was $\Omega = [0, 24] \times [0, 20]$ to ensure that the boundaries were sufficiently far away to avoid interferences from reflections during the simulated time interval. To test the accuracy of the simulation, five probing points were chosen around the obstacle. The points were located in different directions with respect to the obstacle P and the source S. The incident and the reflected acoustic wave passed through these probes at different points in time. This intends to address both phase and amplitude errors that arise from the Brinkman penalization. The porosity was set to $\phi = 1.0$. The viscous and thermal permeability η and η_T were defined respectively with the help of the scaling parameter $\beta = 10^{-6}$ according to Equations (33) and (34). Results were obtained solving the compressible Navier–Stokes equations in two dimensions with a spatial scheme order of $O = 8$, i.e., 64 degrees of freedom per element. Cubical elements with an edge length of $dx = 1/64$ were used to discretize the complete domain. The simulation was carried out for a total time of $t_{max} = 10$ s.

The pressure perturbation in the initial condition resulted in the formation of an acoustic wave that propagated cylindrically outwards as depicted in Figure 11a. Eventually, the wave impinged on the obstacle, where it was reflected as shown in Figure 11b with the pressure perturbation at $t = 4$. The quality of this reflected wave was completely dependent on the quality of the obstacle representation.

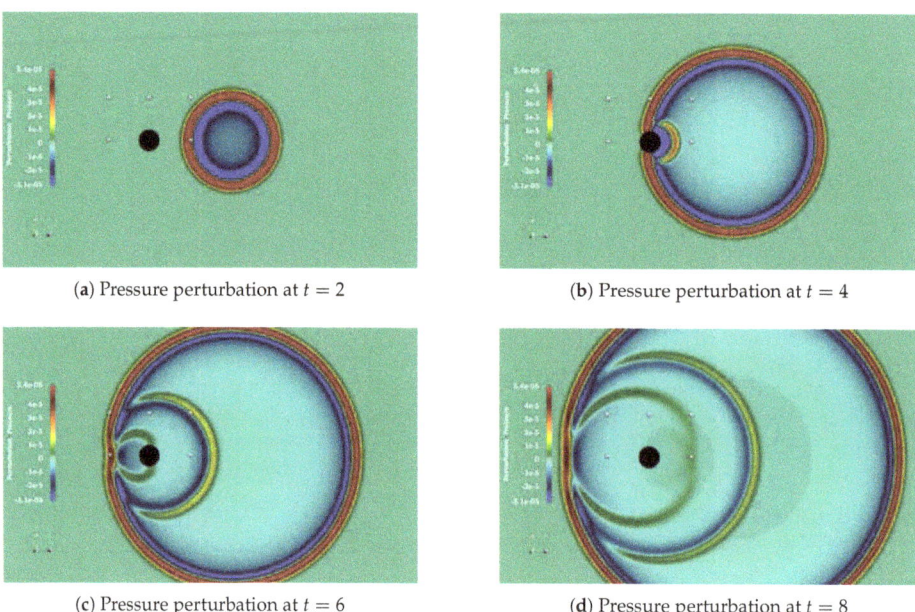

(a) Pressure perturbation at $t = 2$
(b) Pressure perturbation at $t = 4$
(c) Pressure perturbation at $t = 6$
(d) Pressure perturbation at $t = 8$

Figure 11. Simulation snapshots of pressure perturbations captured at successive points in time. The cylindrical obstacle is visible as a black disk and the probe points surrounding it as white dots. The scale of the pressure perturbation is kept constant for all snapshots.

A third wave was generated when the initial wave, disrupted by the obstacle, traveled further to the left and joined again. This is visible in Figure 11c, and its further evolution is visible in Figure 11d, which shows the pressure perturbation at $t = 8$. These three circular acoustic waves had different centers (shifted along the x-axis), but coincided left of the obstacle. As can be seen from these illustrations, the expected reflection pattern was nicely generated by the obstacle representation via the penalization. For a more quantitative assessment of the resulting simulation, we looked at the time evolution of the pressure perturbations at the chosen probing points.

Figure 12 shows the time evolution of the pressure fluctuations monitored at each of the five observation points around the cylinder. The numerical results were compared with the analytical solution for linear equations at these points. Here, we can observe the principal wave and the reflected wave arriving at different probing points at different times. We also observed that the computational results obtained showed an excellent agreement with the analytical solution for all the probes. It nearly perfectly predicted all the amplitudes and pressure behavior without showing phase shifts.

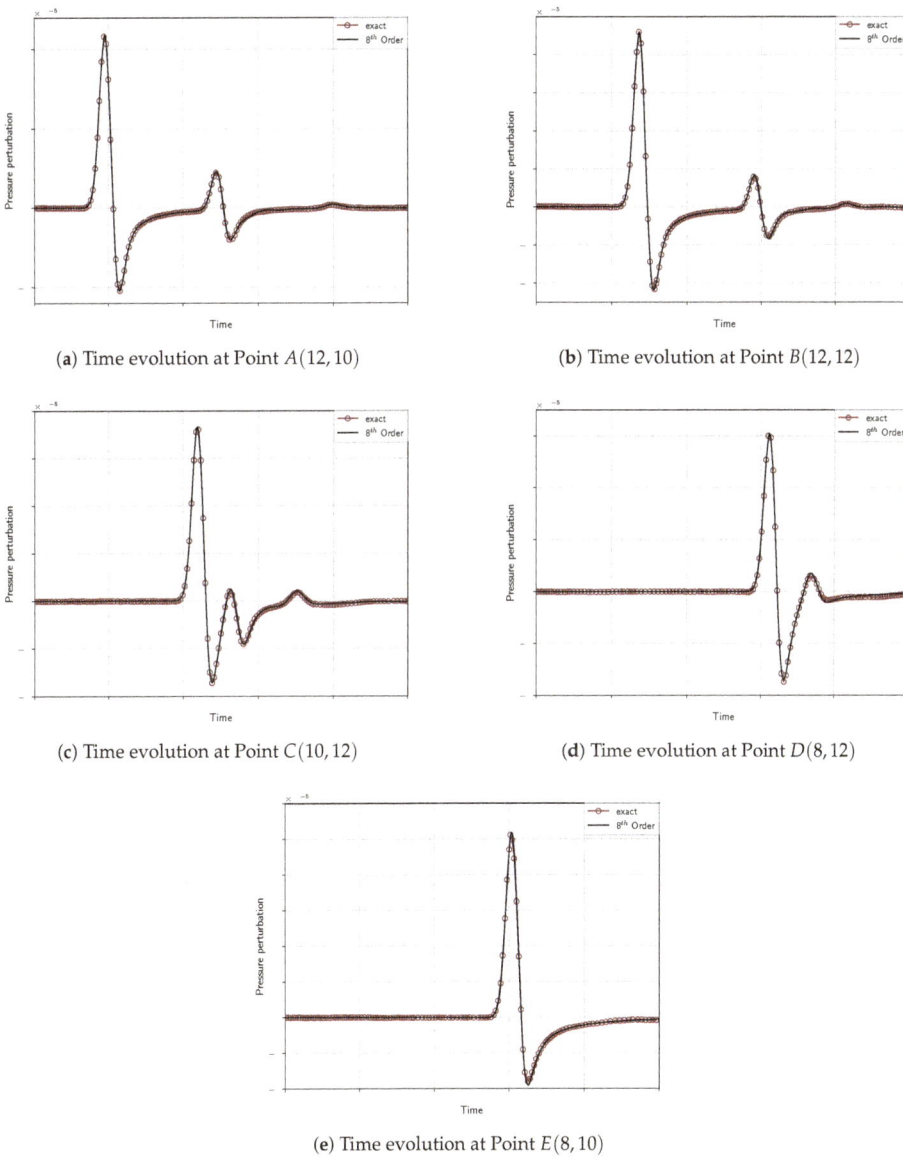

Figure 12. Time evolution of pressure perturbations at all five observation points surrounding the cylinder up to $t = 10$. Be aware that the perturbation pressure plotted along the y axis is scaled differently from probe to probe to illustrate the pressure profile better.

4. Conclusions

We showed that, with the help of an implicit mixed explicit time integration approach, it was feasible to implement wall boundaries accurately in a high-order discontinuous Galerkin scheme. The additional source terms introduced for the penalization can be efficiently computed in the implicit part of the mixed time integration without the need for iterative solvers. This implicit treatment enabled us to utilize arbitrary small values for the permeabilities and freed us from the need for the porosity introduced by Liu and Vasiliyev for compressible flows. The viability of the approach was shown in one-dimensional examples, where we saw that the solid wall can be well approximated with small permeabilities in the high-order discontinuous Galerkin scheme. Even for the reflection of a shock wave, for which a high-order discretization is problematic due to the oscillations incurred by the discontinuity, the penalization provided small errors and convergence with higher polynomial degrees. The real strength of the penalization method, however, came through in multiple dimensions, where curved boundaries could easily be represented by the penalization consistent with the scheme. As an example of such a setting, we looked at the acoustic wave scattering at a cylinder.

With the presented method, it, therefore, was possible to exploit the benefit of reduced memory consumption by the high-order discretization even for complex geometries without the need for advanced mesh generation.

Author Contributions: Conceptualization by N.A. and H.K.; N.A. wrote the original draft preparation and N.E.P. contributed the shock reflection setup; all authors were involved in the review and editing process; N.A., N.E.P. and H.K. worked on the presented methodology; H.K. and N.E.P. worked on the employed software; investigation and validation was carried out by N.A. and N.E.P., they also did the visualization to produce the graphs and images. supervision, S.R.; funding acquisition, S.R.

Funding: Neda Ebrahimi Pour was financially supported by the priority program 1648–Software for Exascale Computing 214 (www.sppexa.de) of the Deutsche Forschungsgemeinschaft.

Conflicts of Interest: The authors declare no conflict of interest. The funders had no role in the design of the study; in the collection, analyses, or interpretation of data; in the writing of the manuscript, or in the decision to publish the results.

References

1. Thompson, J.F.; Warsi, Z.U.; Mastin, C.W. Boundary-fitted coordinate systems for numerical solution of partial differential equations A review. *J. Comput. Phys.* **1982**, *47*, 1–108. [CrossRef]
2. Mittal, R.; Iaccarino, G. Immersed Boundary Methods. *Annu. Rev. Fluid Mech.* **2005**, *37*, 239–261. [CrossRef]
3. Peskin, C.S. Flow patterns around heart valves: A numerical method. *J. Comput. Phys.* **1972**, *10*, 252–271. [CrossRef]
4. Saiki, E.; Biringen, S. Numerical Simulation of a Cylinder in Uniform Flow: Application of a Virtual Boundary Method. *J. Comput. Phys.* **1996**, *123*, 450–465. [CrossRef]
5. Brown-Dymkoski, E.; Kasimov, N.; Vasilyev, O.V. A characteristic based volume penalization method for general evolution problems applied to compressible viscous flows. *J. Comput. Phys.* **2014**, *262*, 344–357. [CrossRef]
6. Arquis, E.; Caltagirone, J.P. Sur les conditions hydrodynamiques au voisinage d'une interface milieu fluide-milieu poreux: Application a' la convection naturelle. *CR Acad. Sci. Paris II* **1984**, *299*, 1–4.
7. Angot, P.; Bruneau, C.H.; Fabrie, P. A penalization method to take into account obstacles in incompressible viscous flows. *Numer. Math.* **1999**, *81*, 497–520. [CrossRef]
8. Kevlahan, N.K.R.; Ghidaglia, J.M. Computation of turbulent flow past an array of cylinders using a spectral method with Brinkman penalization. *Eur. J. Mech. B/Fluids* **2001**, *20*, 333–350. [CrossRef]
9. Liu, Q.; Vasilyev, O.V. A Brinkman penalization method for compressible flows in complex geometries. *J. Comput. Phys.* **2007**, *227*, 946–966. [CrossRef]
10. Jause-Labert, C.; Godeferd, F.; Favier, B. Numerical validation of the volume penalization method in three-dimensional pseudo-spectral simulations. *Comput. Fluids* **2012**, *67*, 41–56. [CrossRef]
11. Pasquetti, R.; Bwemba, R.; Cousin, L. A pseudo-penalization method for high Reynolds number unsteady flows. *Appl. Numer. Math.* **2008**, *58*, 946–954. [CrossRef]

12. Ramière, I.; Angot, P.; Belliard, M. A fictitious domain approach with spread interface for elliptic problems with general boundary conditions. *Comput. Methods Appl. Mech. Eng.* **2007**, *196*, 766–781. [CrossRef]
13. Simulationstechnik und Wissenschaftliches Rechnen Uni Siegen. Ateles Source Code. 2019. Available online: https://osdn.net/projects/apes/scm/hg/ateles/ (accessed on 26 August 2019).
14. Alexander, R. Diagonally Implicit Runge–Kutta Methods for Stiff O.D.E.'s. *SIAM J. Numer. Anal.* **1977**, *14*, 1006–1021. [CrossRef]
15. Zudrop, J. Efficient Numerical Methods for Fluid- and Electrodynamics on Massively Parallel Systems. Ph.D. Thesis, RWTH Aachen University, Aachen, Germany, 2015.
16. Hesthaven, J.S.; Warburton, T. *Nodal Discontinuous Galerkin Methods: Algorithms, Analysis, and Applications*, 1st ed.; Springer: New York, NY, USA, 2007.
17. Piquet, A.; Roussel, O.; Hadjadj, A. A comparative study of Brinkman penalization and direct-forcing immersed boundary methods for compressible viscous flows. *Comput. Fluids* **2016**, *136*, 272–284. [CrossRef]
18. Ben-Dor, G.; Igra, O.; Elperin, T. (Eds.) *Handbook of Shock Waves*; Academic Press: Cambridge, MA, USA, 2001.
19. Glazer, E.; Sadot, O.; Hadjadj, A.; Chaudhuri, A. Velocity scaling of a shock wave reflected off a circular cylinder. *Phys. Rev. E* **2011**, *83*, 066317. [CrossRef]
20. Tam, C.K.W.; Hardin, J.C. *Second Computational Aeroacoustics (CAA) Workshop on Benchmark Problems*; NASA, Langley Research Center: Hampton, VA, USA, 1997.

© 2019 by the authors. Licensee MDPI, Basel, Switzerland. This article is an open access article distributed under the terms and conditions of the Creative Commons Attribution (CC BY) license (http://creativecommons.org/licenses/by/4.0/).

Article

Numerical Method for Dirichlet Problem with Degeneration of the Solution on the Entire Boundary

Viktor A. Rukavishnikov *,† and **Elena I. Rukavishnikova** †

Computing Center of Far-Eastern Branch, Russian Academy of Sciences, Kim-Yu-Chen Str. 65, Khabarovsk 680000, Russia; rukavishnikova-55@mail.ru
* Correspondence: vark0102@mail.ru; Tel.: +8-421-257-2620
† These authors contributed equally to this work.

Received: 10 November 2019; Accepted: 23 November 2019; Published: 26 November 2019

Abstract: The finite element method (FEM) with a special graded mesh is constructed for the Dirichlet boundary value problem with degeneration of the solution on the entire boundary of the two-dimensional domain. A comparative numerical analysis is performed for the proposed method and the classical finite element method for a set of model problems in symmetric domain. Experimental confirmation of theoretical estimates of accuracy is obtained and conclusions are made.

Keywords: boundary value problems with degeneration of the solution on entire boundary of the domain; the method of finite elements; special graded mesh

1. Introduction

As is known, classical solutions for boundary value problems for elliptic equations with discontinuous coefficients do not exist. Therefore, the notion of a generalized (weak) solution was introduced. Based on this definition and on the Galerkin method, numerous numerical methods were developed for finding approximate solutions of such problems. However, these methods for boundary value problems with singularity lose accuracy, which depends on the smoothness of the solution of the original differential problem (see, for example, [1,2]). The singularity of the solution of the boundary value problem can be caused by the presence of re-entrant corners on the domain boundary, by the degeneration of the coefficients and right-hand sides of equation and boundary conditions, or by the internal properties of the solution (see, for example, [3–9]). For boundary problems with a singularity, we proposed to determine an R_ν-generalized solution. The existence, uniqueness and differential properties of this kind of solution in the weighted Sobolev spaces were studied in [10–15]. Based the R_ν-generalized solution, a weighted finite element method was developed for boundary value problems for elliptic equations in two-dimensional domain with a singularity in a finite set of boundary points [16–20]. A weighted FEM was constructed and studied for the Lamé system in a domain with re-entrant corners [21,22]. To find an approximate solution of Maxwell's equations in an L-shaped domain, a weighted edge-based finite element method was proposed in [23,24]. In [25,26] a weight analogue of the condition of Ladyzhenskaya-Babuška-Brezzi was proved, a numerical method was developed for the Stokes and Oseen problems in domains with corner singularity. The main feature of all the developed methods is the convergence of the approximate solution to the exact one with the rate $O(h)$ in the norms of the Sobolev and Monk weighted spaces, regardless of the reasons causing the solution singularity and its value.

In this paper we consider the Dirichlet problem for an elliptic equation with degeneration of the solution on the entire boundary of a two-dimensional domain. In [27] a finite element method was constructed for this problem and the convergence of this method was established. The paper [28] singles out the weighted subspace of functions for which the approximate solutions converge to an exact solution with a speed $O(h)$ on a mesh with a special compression of nodes close to the boundary (see [29]). The compression parameters depend on the constructed subspace. Our method of constructing mesh with a special compression of nodes differs from the methods proposed by other authors (see, for example, [30–32]).

Here we test model problems with singularities in a symmetric domain. We carry out a comparative numerical analysis of finite element methods on quasi-uniform meshes and meshes with a special compression of nodes close to the boundary. We obtain experimental confirmation of theoretical estimates and demonstrate the advantage of the proposed method over the classical finite element method. By analogy with [22], we found that it is impossible to use FEM with a strong thickening of mesh, and introduction of an R-generalized solution is required. The existence and uniqueness of the R-generalized solution for this problem were proved in [33].

2. Problem Formulation

Let Ω be a bounded convex two-dimensional domain with twice differentiable boundary $\partial\Omega$, and let $\overline{\Omega}$ be the closure of Ω, i.e., $\overline{\Omega} = \Omega \cup \partial\Omega$; $x = (x_1, x_2)$ and $dx = dx_1 dx_2$.

We assume that a positive function $\rho(x)$ belongs to the space $C^{(2)}(\Omega)$ and coincides in the boundary strip of width $d > 0$ with the distance from x ($x \in \Omega$) to the boundary $\partial\Omega$.

We introduce the weighted Sobolev space $W_{2,\eta}^s(\Omega)$ with the norm

$$\|v\|_{W_{2,\eta}^s(\Omega)} = \|v\|_{L_2(\Omega)} + \sum_{\substack{m_1, m_2 = 0 \\ |m|=s}} \left\| \rho^{-\eta} \frac{\partial^{|m|} v}{\partial x_1^{m_1} \partial x_2^{m_2}} \right\|_{L_2(\Omega)},$$

where η is a real number satisfying the inequalities $\frac{1}{2} - s < \eta < \frac{1}{2}$; $s = 1, 2$; $m = (m_1, m_2)$, $|m| = m_1 + m_2$, m_1, m_2 are integer nonnegative numbers.

Let

$$\mathring{W}_{2,\eta}^s(\Omega) = \{v: v \in W_{2,\eta}^s(\Omega), \ v|_{\partial\Omega} = 0\}.$$

We denote by $L_{2,-1-\eta}(\Omega)$ the space of functions f with the norm

$$\|f\|_{L_{2,-1-\eta}(\Omega)} = \left(\int_{\Omega} |\rho^{1+\eta} f|^2 \, dx \right)^{1/2}.$$

We consider the first boundary value problem for a second order elliptic equation

$$-\sum_{k=1}^{2} \frac{\partial}{\partial x_k} \left(a_{kk}(x) \frac{\partial u}{\partial x_k} \right) + a(x) u = f \quad \text{in } \Omega,$$

$$u = 0 \quad \text{on } \partial\Omega.$$

(1)

We suppose that the input data of Equation (1) satisfy the conditions:

(a)
$$f \in L_{2,-1-\alpha}(\Omega), \qquad (2)$$

(b) $a_{kk}(x)$ ($k = 1, 2$) are differentiable functions on Ω, such that the inequalities

$$|a_{kk}(x)| \leq C_1 \rho^{-2\alpha}(x), \qquad (3)$$

$$\left| \frac{\partial a_{kk}(x)}{\partial x_1} \right|, \left| \frac{\partial a_{kk}(x)}{\partial x_2} \right| \leq C_2 \rho^{-2\alpha - 1}(x), \qquad (4)$$

$$\sum_{k=1}^{2} a_{kk}(x)\xi_k^2 \geq C_3 \rho^{-2\alpha}(x) \sum_{k=1}^{2} \xi_k^2, \quad x \in \Omega, \quad C_3 > 0, \qquad (5)$$

hold,
(c) the function $a(x)$ satisfies the inequalities

$$0 < a(x) \leq C_4 \rho^{-2\alpha-2}(x), \quad x \in \Omega. \qquad (6)$$

Here C_i, $(i = 1, \ldots, 4)$ are constants independent of x, ξ_1 and ξ_2 are any real parameters, $\alpha \in \left(-\frac{1}{2}, \frac{1}{2}\right)$.

Remark 1. *If Conditions (2)–(6) are fulfilled for the input data, Equation (1) is called a Dirichlet boundary value problem for an elliptic equation with degeneration of the solution on the entire boundary of a two-dimensional domain. Such problems are encountered in gas dynamics, electromagnetism and other subject areas of mathematical physics. The differential properties of solutions of problems with degeneracy on the entire boundary were studied, for the first time, in [7–9].*

We introduce the bilinear and linear forms

$$E(u,w) = \int_\Omega \left(\sum_{k=1}^{2} a_{kk}(x) \frac{\partial u}{\partial x_k} \frac{\partial w}{\partial x_k} + a(x) u w\right) dx, \qquad (f,w) = \int_\Omega f w\, dx.$$

A function u in $\mathring{W}^1_{2,\alpha}(\Omega)$ is called a generalized solution of the first boundary value Equation (1) if for any w in $\mathring{W}^1_{2,\alpha}(\Omega)$ the identity

$$E(u,w) = (f,w)$$

holds.

We note that if Conditions (2)–(6) are satisfied, then there exists a unique generalized solution of the Equation (1) in the space $\mathring{W}^1_{2,\alpha}(\Omega)$ (see Theorem 1 from [8]). In addition $u \in \mathring{W}^2_{2,\alpha-1}(\Omega)$ (see Theorem 1 from [9]). Moreover, if the function $f \in L_{2,-1-\alpha+\lambda}(\Omega) \subset L_{2,-1-\alpha}(\Omega)$ $\left(-\frac{1}{2} < \alpha < \alpha + \lambda < \frac{1}{2}\right)$ and the parameter λ is sufficiently small, then the generalized solution u belongs to the space $\mathring{W}^2_{2,\alpha+\lambda-1}(\Omega)$ which is a subspace of $\mathring{W}^2_{2,\alpha-1}(\Omega)$ (see [28]).

Remark 2. *Knowing that the solution belongs to the space $\mathring{W}^2_{2,\alpha+\lambda-1}(\Omega)$ allows us to construct a finite element method for finding a generalized solution for the Dirichlet problem with the degeneration of the solution on the entire boundary of the domain with a convergence speed $O(h)$ in the norm $W^1_{2,\alpha}(\Omega)$.*

3. The Scheme of the Finite Element Method

We construct a scheme of the finite element method for finding an approximate generalized solution of the first boundary value Equation (1). We perform a triangulation of the domain Ω (see, for example, Figure 1).

We draw the curves Γ_j, $j = 0, \ldots, n$, at distance $b\left(\frac{j}{n}\right)^r$, $j = 0, \ldots, n$, to the boundary $\partial \Omega$. Here r is the exponent of compression and $r > 1$; $0 < b < \frac{\delta_\Omega}{2} \leq d$, δ_Ω is the diameter of the circle inscribed in Ω. In this case the line Γ_n divides the domain Ω into two subdomains Ω_1 and Ω_2. The subdomain Ω_1 is the outer domain on the boundary strip of width b, Ω_2 is the inner domain. On each curve Γ_j, $j = 0, \ldots, n$, $(\Gamma_0 = \partial\Omega, \Gamma_n = \partial\Omega_1)$ we fix M_j equidistant points, which we call the nodes. Here $M_j = \left[l_j / \left(b\left(\left(\frac{j}{n}\right)^r - \left(\frac{j-1}{n}\right)^r\right)\right)\right] + 1$, $j = 1, \ldots, n$, l_j is the length of the curve Γ_j ($[x]$ denotes the integer part of x) and $M_0 = 2M_1$. All nodes on the curve Γ_j, $j = 0, \ldots, n$, are connected by the broken line. Then, we connect each node on the curve Γ_{j-1}, $j = 1, \ldots, n$, with closest nodes on the curve Γ_j. As a result, the subdomain Ω_1 is divided into triangles with the compression of nodes to the boundary $\partial\Omega$. The union of all triangles with vertices on Γ_{j-1} and Γ_j is a layer Q_j^h. (In Figure 1 the subdomain Ω_1 is divided into the layers $Q_1^h, \ldots, Q_4^h, \Gamma_4 = \partial\Omega_1$). The parameter h denotes the greatest in length of the sides of the triangles in Q_n^h.

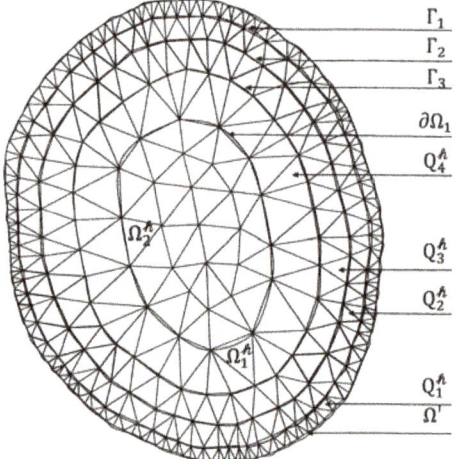

Figure 1. Triangulation of domain Ω.

The subdomain Ω_2 is divided quasi-uniformly into a finite number of the triangles. The sides of these triangles can not be greater than h. Moreover, the vertices of the triangles on the boundary $\partial\Omega_1$ belong to the set of vertices of the triangles in $\partial\Omega_2$.

The algorithm and code description of this triangulation are given in [29].

Let Ω^h be the union of closed triangles $\{K\} = \{K_1, \ldots, K_{N_h}\}$, and K_j, $j = 1, \ldots, N_h$, is the finite element. The vertices G_i, $i = 1, \ldots, N$, of these triangles are the nodes of the triangulation. We denote by N' the number of the internal nodes. To each node G_i, $i = 1, \ldots, N'$, we assign the function $\varphi_i(x)$ which is equal to 1 at the point G_i and zero at all other nodes, and $\varphi_i(x)$ is linear on each triangle K. We denote by V^h the linear span $\{\varphi_i\}_{i=1}^{N'}$. Next, we associate the following discrete problem with the constructed finite-dimensional space $V^h \subset \mathring{W}_{2,\alpha}^1(\Omega)$: find the function $u_h \in V^h$ satisfying the equality

$$E(u_h, w_h) = (f, w_h)$$

for any function $w_h \in V^h$.

An approximate (finite element) generalized solution will be found in the form

$$u_h(x) = \sum_{i=1}^{N'} a_i \varphi_i(x),$$

where $a_i = u_h(G_i)$. We assume that $u_h(x) = 0$, $x \in \Omega \setminus \Omega^h$.

The coefficients a_i are defined from system of equations

$$E(u_h, \varphi_i) = (f, \varphi_i), \quad i = 1, \ldots, N'$$

or

$$\hat{A}a = F,$$

where

$$a = (a_1, \ldots, a_{N'})^T, \quad F = (F_1, \ldots, F_{N'})^T, \quad \hat{A} = (A_{ij}),$$

$$A_{ij} = E(\varphi_i, \varphi_j), \quad F_i = (f, \varphi_i), \quad i, j = 1, \ldots, N'.$$

It is obvious that the approximate generalized solution of Problems (1)–(6) by the finite element method exists and is unique.

For the performed triangulation of the domain Ω with the exponent of the compression of the mesh $r = \frac{1}{\lambda}$ and for functions in the space $\mathring{W}^2_{2,\alpha+\lambda-1}(\Omega)$ we have convergence estimates:

$$\|u - u_h\|_{W^1_{2,\alpha}(\Omega)} \leq C_5 h \|f\rho^{1+\alpha-\lambda}\|_{L_2(\Omega)},$$

$$\|u - u_h\|_{L_2(\Omega)} \leq C_6 h^2 \|f\rho^{1+\alpha-\lambda}\|_{L_2(\Omega)}.$$

Here, the positive constants C_5, C_6 are independent of u, u_h, f and h.

4. Numerical Experiments

In this section we present the results of numerical experiments for two model problems.

Let Ω be a circle of unit radius with center at the point $(2,2)$. We consider the boundary value Equation (1) in the domain Ω. The right-hand side and coefficients of Equation (1) are given as

$$f(x) = 4(1 + \mu - t)\Big((\mu - 2\alpha - t)\tilde{\rho}^{\mu-2\alpha-t-1}(x)(1 - \tilde{\rho}(x)) - \tilde{\rho}^{\mu-2\alpha-t}(x)\Big) + \tilde{\rho}^{1+\mu-t}(x),$$

$$a_{11}(x) = a_{22}(x) = \tilde{\rho}^{-2\alpha}(x), \quad a(x) = 1, \quad \alpha \in \left(-\frac{1}{2}, \frac{1}{2}\right), \quad \mu \in \left(\alpha, \frac{1}{2}\right), \quad t < \frac{1}{2},$$

where $\tilde{\rho}(x) = 1 - (x_1 - 2)^2 - (x_2 - 2)^2$ and $\tilde{\rho}(x)$ be a function that is infinitely differentiable and satisfies the following conditions:

$C_7 \rho(x) \leq \tilde{\rho}(x) \leq C_8 \rho(x)$ for each point x of the domain Ω; $0 < C_7 \leq C_8 < \infty$.

The exact solution of this problem is $u(x) = \tilde{\rho}^{1-\mu-t}(x)$.

For finding an approximate solution of model problems we used mesh with the compression of nodes to the boundary (R_c), quasi-uniform mesh (R_q) and the finite element method scheme from paragraph three. For the mesh R_c we set the number of layers n and the exponent of compression of the mesh $r = \frac{1}{\tau}$, $\tau = \mu - \alpha$.

We investigate the convergence rate of the approximate solution u_h to the exact one in the norms of the spaces $L_2(\Omega)$ and $W^1_{2,\alpha}(\Omega)$ on the mesh R_c and R_q. The absolute value of the error $e = |u - u_h|$ in the mesh nodes G_i on the mesh R_c and R_q is analyzed.

Model Problem 1. We set the parameters $\alpha = 0.01$, $\mu = 0.49$, $t = 0.499$, at which the solution, the coefficients and the right-hand side of the equation in Equation (1) have the form

$$u(x) = \tilde{\rho}^{0.991}(x),$$

$$a_{11}(x) = a_{22}(x) = \tilde{\rho}^{-0.02}(x), \quad a(x) = 1,$$

$$f(x) = 0.114956 \cdot \tilde{\rho}^{-1.029}(x) + 3.849044 \cdot \tilde{\rho}^{-0.029}(x) + \tilde{\rho}^{0.991}(x),$$

the exponent of compression of the mesh $r = 2.08(3)$.

In Table 1 we give the number of nodes and their percentage to the total number of mesh nodes N, in which the absolute value of the error $e = |u - u_h|$ is not less than the specified value of the limit error. In this Table the patterns of the absolute error distribution at the nodes of the R_q and R_c meshes are also showed. We present data on the R_c mesh for N nodes and on the R_q mesh for N and $\frac{N}{2}$ nodes.

In Table 2 we present the norms of the difference between an exact and an approximate solution $\|e\|_{L_2(\Omega)} = \|u - u_h\|_{L_2(\Omega)}$ and $\|e\|_{W^1_{2,\alpha}(\Omega)} = \|u - u_h\|_{W^1_{2,\alpha}(\Omega)}$ for R_q and R_c and find the ratios of the norms β when the mesh parameter h is reduced by a factor two. The value of the parameter h in the domain Ω_2 for R_c varies by changing number of layers n.

Table 1. Absolute value of the error e for Model Problem 1.

Quasi-Uniform Mesh (R_q)		Absolute Error Distribution	Specified Limited Error	Percent	Number of Nodes
Number of nodes N	10,849,474		■ $e \geq 3 \times 10^{-6}$	0.00%	69
			▨ $1 \times 10^{-6} \leq e < 3 \times 10^{-6}$	89.37%	9,696,362
			▨ $7 \times 10^{-7} \leq e < 1 \times 10^{-6}$	10.61%	1,151,232
			▨ $3 \times 10^{-7} \leq e < 7 \times 10^{-7}$	0.01%	1196
h	0.00055		▨ $1 \times 10^{-7} \leq e < 3 \times 10^{-7}$	0.00%	372
			□ $0 \leq e < 1 \times 10^{-7}$	0.00%	243

Refined Mesh (R_c)		Absolute Error Distribution	Specified Limited Error	Percent	Number of Nodes
N	10,755,478		■ $e \geq 3 \times 10^{-6}$	0.00%	0
Number of nodes in domain Ω_2	10,661,162		▨ $1 \times 10^{-6} \leq e < 3 \times 10^{-6}$	0.00%	0
			▨ $7 \times 10^{-7} \leq e < 1 \times 10^{-6}$	0.00%	4
h	0.000556		▨ $3 \times 10^{-7} \leq e < 7 \times 10^{-7}$	44.46%	4,781,879
n	3		▨ $1 \times 10^{-7} \leq e < 3 \times 10^{-7}$	55.44%	5,962,761
b	1/1024		□ $0 \leq e < 1 \times 10^{-7}$	0.10%	10,834

Refined Mesh (R_c)		Absolute Error Distribution	Specified Limited Error	Percent	Number of Nodes
N	4,974,486		■ $e \geq 3 \times 10^{-6}$	0.00%	0
Number of nodes in domain Ω_2	4,241,164		▨ $1 \times 10^{-6} \leq e < 3 \times 10^{-6}$	0.00%	0
			▨ $7 \times 10^{-7} \leq e < 1 \times 10^{-6}$	0.00%	0
h	0.00087		▨ $3 \times 10^{-7} \leq e < 7 \times 10^{-7}$	2.75%	136,821
n	18		▨ $1 \times 10^{-7} \leq e < 3 \times 10^{-7}$	83.18%	4,137,684
b	1/128		□ $0 \leq e < 1 \times 10^{-7}$	14.07%	699,981

Table 2. The errors $\|e\|_{L_2(\Omega)}$ and $\|e\|_{W^1_{2,a}(\Omega)}$ for meshes R_q and R_c for Model Problem 1.

Quasi-Uniform Mesh (R_q)					Refined Mesh (R_c), $b = 1/128$				
h	$\|e\|_{L_2(\Omega)}$	β	$\|e\|_{W^1_{2,a}(\Omega)}$	β	h	$\|e\|_{L_2(\Omega)}$	β	$\|e\|_{W^1_{2,a}(\Omega)}$	β
0.0022	5.40×10^{-6}		3.83×10^{-3}		0.0035	3.00×10^{-6}		5.11×10^{-3}	
		1.68		1.85			4.21		2.06
0.0011	3.21×10^{-6}		2.22×10^{-3}		0.00169	7.12×10^{-7}		2.48×10^{-3}	
		1.83		1.72			4.11		2.03
0.00055	1.75×10^{-6}		1.36×10^{-3}		0.00083	1.73×10^{-7}		1.23×10^{-3}	

The distribution of the absolute values of the error e in the mesh nodes with a decrease in the h parameter by a factor of two on the meshes R_q and R_c is given in Table 3.

Table 3. The distribution of the error e on the grids R_q and R_c as h changes for Model Problem 1.

Specified Limited Error	Quasi-Uniform Mesh (R_q)		
■ $e \geq 3 \times 10^{-6}$			
■ $1 \times 10^{-6} \leq e < 3 \times 10^{-6}$			
■ $7 \times 10^{-7} \leq e < 1 \times 10^{-6}$			
■ $3 \times 10^{-7} \leq e < 7 \times 10^{-7}$			
■ $1 \times 10^{-7} \leq e < 3 \times 10^{-7}$			
□ $0 \leq e < 1 \times 10^{-7}$			
N	338,449	1,355,498	5,427,739
h	0.0031	0.0015	0.00078
Specified Limited Error	Refined Mesh (R_c), $b = 1/128$		
	$n = 5$	$n = 10$	$n = 20$
■ $e \geq 3 \times 10^{-6}$			
■ $1 \times 10^{-6} \leq e < 3 \times 10^{-6}$			
■ $7 \times 10^{-7} \leq e < 1 \times 10^{-6}$			
■ $3 \times 10^{-7} \leq e < 7 \times 10^{-7}$			
■ $1 \times 10^{-7} \leq e < 3 \times 10^{-7}$			
□ $0 \leq e < 1 \times 10^{-7}$			
N	425,760	1,569,052	6,129,755
Number of nodes in domain Ω_2	386,628	1,375,684	5,200,079
h in domain Ω_2	0.0029	0.0015	0.00079

Model Problem 2. We set the parameters $\alpha = -0.49$, $\mu = 0.01$, $t = 0.49$, at which the solution, the coefficients and the right-hand side of Equation (1) have the form

$$u(x) = \tilde{\rho}^{0.52}(x),$$

$$a_{11}(x) = a_{22}(x) = \tilde{\rho}^{0.98}(x), \quad a(x) = 1,$$

$$f(x) = -1.04 \cdot \tilde{\rho}^{-0.5}(x) + 3.12 \cdot \tilde{\rho}^{0.5}(x) + \tilde{\rho}^{0.52}(x),$$

the exponent of compression of the mesh $r = 2$.

A numerical analysis of this problem was carried out by analogy with Model Problem 1. The results of the research are presented in Tables 4–6.

Table 4. Absolute value of the error e for Model Problem 2.

Quasi-Uniform Mesh (R_q)		Absolute Error Distribution	Specified Limited Error	Percent	Number of Nodes
N	2,713,152		$e \geq 5 \times 10^{-3}$	50.40%	1,367,499
			$2 \times 10^{-3} \leq e < 5 \times 10^{-3}$	49.60%	1,345,653
			$1 \times 10^{-3} \leq e < 2 \times 10^{-3}$	0.00%	0
h	0.00055		$7 \times 10^{-4} \leq e < 1 \times 10^{-3}$	0.00%	0
			$4 \times 10^{-4} \leq e < 7 \times 10^{-4}$	0.00%	0

Refined Mesh (R_c)		Absolute Error Distribution	Specified Limited Error	Percent	Number of Nodes
N	3,254,432		$e \geq 5 \times 10^{-3}$	0.00%	0
Number of nodes in domain Ω_2	2,484,744		$2 \times 10^{-3} \leq e < 5 \times 10^{-3}$	0.00%	0
h in domain Ω_2	0.0011		$1 \times 10^{-3} \leq e < 2 \times 10^{-3}$	26.32%	856,521
n	27		$7 \times 10^{-4} \leq e < 1 \times 10^{-3}$	31.94%	1,039,571
b	1/64		$4 \times 10^{-4} \leq e < 7 \times 10^{-4}$	41.74%	1,358,340

Table 5. The error $\|e\|_{W^1_{2,\alpha}(\Omega)}$ for meshes R_q and R_c for Model Problem 2.

Quasi-Uniform Mesh (R_q)				Refined Mesh (R_c), $b = 1/64$			
h	$\|e\|_{W^1_{2,\alpha}(\Omega)}$	β	n	h in domain Ω_2	$\|e\|_{W^1_{2,\alpha}(\Omega)}$	β	
0.0022	0.065816		3	0.0068	0.04562		
		1.43				2.08	
0.0011	0.046016		8	0.0032	0.021939		
		1.43				2.00	
0.00055	0.032220		18	0.0016	0.010989		

Table 6. The distribution of the error e on the grids R_q and R_c as h changes for Model Problem 2.

Specified Limited Error	Quasi-Uniform Mesh (R_q)		
$e \geq 5 \times 10^{-3}$			
$2 \times 10^{-3} \leq e < 5 \times 10^{-3}$			
$1 \times 10^{-3} \leq e < 2 \times 10^{-3}$			
$7 \times 10^{-4} \leq e < 1 \times 10^{-3}$			
$4 \times 10^{-4} \leq e < 7 \times 10^{-4}$			

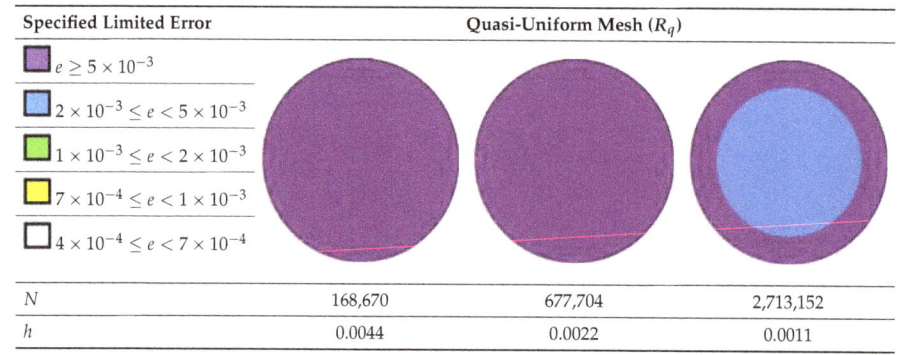

N	168,670	677,704	2,713,152
h	0.0044	0.0022	0.0011

Table 6. Cont.

Specified Limited Error	Refined Mesh (R_c)		
	$n = 6$	$n = 12$	$n = 24$
■ $e \geq 5 \times 10^{-3}$			
■ $2 \times 10^{-3} \leq e < 5 \times 10^{-3}$			
■ $1 \times 10^{-3} \leq e < 2 \times 10^{-3}$			
□ $7 \times 10^{-4} \leq e < 1 \times 10^{-3}$			
□ $4 \times 10^{-4} \leq e < 7 \times 10^{-4}$			
N	166,751	643,498	2,567,442
Number of nodes in domain Ω_2	139,635	514,952	1,972,944
h in domain Ω_2	0.0050	0.0025	0.00127

In Figure 2a,b we present graphs of the error $\|e\|_{W_{2,a}^1(\Omega)} = \|u - u_h\|_{W_{2,a}^1(\Omega)}$ as a function of the parameter h on the grids R_q and R_c in a logarithmic scale. In the first case the parameter h decreases due to an increase in the number of layers n at a fixed value $b = 1/64$ (Figure 2a). In the second case h changes due to a decrease in the width of the border strip b at a fixed number n (Figure 2b).

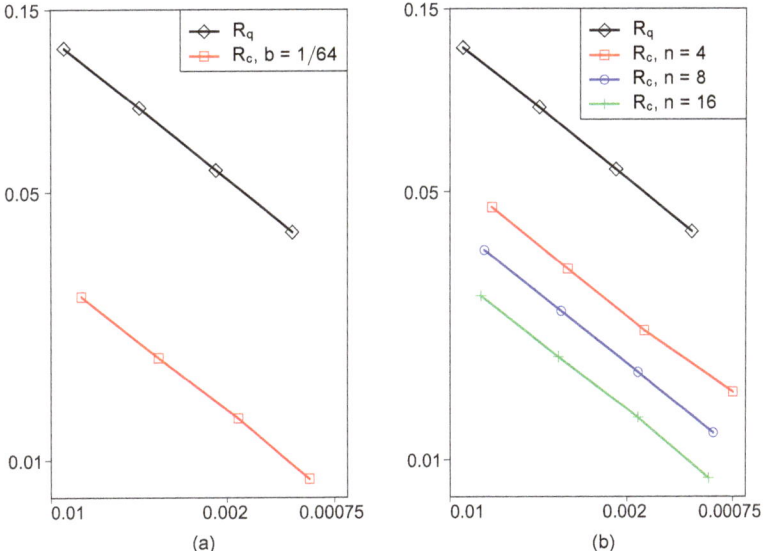

Figure 2. The graph of the error $\|e\|_{W_{2,a}^1(\Omega)}$ on the grids R_q and R_c as h changes in a logarithmic scale for Model Problem 2. For R_c (**a**) $b = const = 1/64$, n is a variable; (**b**) n is a constant, b is a variable.

5. Conclusions

We can conclude according to the results of numerical experiments:

- An approximate generalized solution of Equation (1) on grids with an appropriate compression of nodes (close) to the boundary of the domain converges to an exact solution with a speed $O(h^2)$ in the norm of the space $L_2(\Omega)$ and $O(h)$ in the norm of the space $W_{2,a}^1(\Omega)$ (see Tables 2 and 5);

- the absolute value of the error is an order of magnitude smaller on the mesh with the compression of nodes to the boundary of the domain and with an optimal choice of parameters b and n than on a quasi-uniform mesh (see Tables 1, 3, 4 and 6);
- to reduce the absolute value of the error it is more expedient to increase the number of layers n than to reduce the width of the boundary ring domain; in this case the absolute value of the error decreases faster;
- for meshes of large dimensionality it is advisable to use the weighted finite element method.

In the next papers we plan to develop the proposed finite element method for boundary value problems with inhomogeneous boundary conditions for self-adjoint differential equations of second and higher orders with weaker conditions on the input data of the problem, in particular

$$a(x) \geq -c, \quad x \in \Omega, \quad c \text{ is finite constant.}$$

Author Contributions: V.A.R. and E.I.R. contributed equally in each stage of the work. All authors read and approved the final version of the paper.

Funding: This research was funded by RFBR grant number 20-01-00022.

Acknowledgments: This research was supported in through computational resources provided by the Shared Facility Center "Data Center of FEB RAS".

Conflicts of Interest: The authors declare no conflict of interest.

References

1. Ciarlet, P. *The Finite Element Method for Elliptic Problems*; Studies in Mathematics and Its Applications, North-Holland: Amsterdam, The Netherlands, 1978.
2. Samarskii, A.A.; Lazarov, R.D.; Makarov, V.L. *Finite Difference Schemes for Differential Equations with Generalized Solutions*; Visshaya Shkola: Moscow, Russia, 1987; 296p.
3. Kondrat'ev, V.A. Boundary problems for elliptic equations in domains with conical or angular points. *Trans. Mosc. Math. Soc.* **1967**, *16*, 227–313.
4. Kondrat'ev, V.A.; Oleinik, O.A. Boundary-value problems for partial differential equations in non-smooth domains. *Rus. Math. Surv.* **1983**, *38*, 1–86, doi:10.1070/rm1983v038n02abeh003470.
5. Maz'ya, V.G. On weak solutions of the Dirichlet and Neumann problems. *Trans. Mosc. Math. Soc.* **1971**, *20*, 135–172.
6. Maz'ya, V.G.; Plamenevskii, B.A. On coeffcients in the asymptotics of solutions of elliptic problems in domains with conical points. *Math. Nachr.* **1977**, *76*, 29–60, doi:10.1002/mana.19770760103.
7. Nikol'skij, S.M. A Variational Problem for an Equation of Elliptic Type with Degeneration on the Boundary. *Proc. Steklov Inst. Math.* **1981**, *150*, 227–254.
8. Lizorkin, P.I.; Nikol'skij, S.M. Elliptic equations with degeneracy. Differential properties of solutions. *Sov. Math. Dokl.* **1981**, *23*, 268–271.
9. Lizorkin, P.I.; Nikol'skij, S.M. Coercive properties of an elliptic equation with degeneracy (the case of generalized solutions). *Sov. Math. Dokl.* **1981**, *24*, 21–23.
10. Rukavishnikov, V.A. The Dirichlet problem with the noncoordinated degeneration of the initial data. *Dokl. Akad. Nauk* **1994**, *337*, 447–449.
11. Rukavishnikov, V.A. The Dirichlet problem for a second-order elliptic equation with noncoordinated degeneration of the input data. *Differ. Equ.* **1996**, *32*, 406–412.
12. Rukavishnikov, V.A. On the uniqueness of the R_ν-generalized solution of boundary value problems with noncoordinated degeneration of the initial data. *Dokl. Math.* **2001**, *63*, 68–70.
13. Rukavishnikov, V.A.; Ereklintsev, A.G. On the coercivity of the R_ν-generalized solution of the first boundary value problem with coordinated degeneration of the input data. *Differ. Equ.* **2005**, *41*, 1757–1767, doi:10.1007/s10625-006-0012-5.
14. Rukavishnikov, V.A.; Kuznetsova, E.V. Coercive estimate for a boundary value problem with noncoordinated degeneration of the data. *Differ. Equ.* **2007**, *43*, 550–560, doi:10.1134/S0012266107040131.

15. Rukavishnikov, V.A. On the existence and uniqueness of an R_ν-generalized solution of a boundary value problem with uncoordinated degeneration of the input data. *Dokl. Math.* **2014**, *90*, 562–564, doi:10.1134/S1064562414060155.
16. Rukavishnikov, V.A. The weight estimation of the speed of difference scheme convergence. *Dokl. Akad. Nauk SSSR* **1986**, *288*, 1058–1062.
17. Rukavishnikov, V.A.; Rukavishnikova, E.I. Finite-Element Method for the 1St Boundary-Value Problem with the Coordinated Degeneration of the Initial Data. *Dokl. Akad. Nauk* **1994**, *338*, 731–733.
18. Rukavishnikov, V.A. Methods of numerical analysis for boundary value problem with strong singularity. *Rus. J. Numer. Anal. Math. Model.* **2009**, *24*, 565–590, doi:10.1016/j.cam.2010.01.020.
19. Rukavishnikov, V.A.; Rukavishnikova, H.I. The finite element method for a boundary value problem with strong singularity. *J. Comput. Appl. Math.* **2010**, *234*, 2870–2882, doi:10.1515/RJNAMM.2009.035.
20. Rukavishnikov, V.A.; Rukavishnikova, H.I. On the error estimation of the finite element method for the boundary value problems with singularity in the Lebesgue weighted space. *Numer. Funct. Anal. Opt.* **2013**, *34*, 1328–1347, doi:10.1080/01630563.2013.809582.
21. Rukavishnikov, V.A.; Nikolaev, S.G. Weighted finite element method for an elasticity problem with singularity. *Dokl. Math.* **2013**, *88*, 705–709, doi:10.1134/S1064562413060215.
22. Rukavishnikov, V.A.; Rukavishnikova, H.I. *Weighted Finite-Element Method for Elasticity Problems with Singularity*, p razvan ed.; Finite Element Method. Simulation, Numerical Analysis and Solution Techniques, IntechOpen Limited: London, UK, 2018; pp. 295–311, doi:10.5772/intechopen.72733.
23. Rukavishnikov, V.A.; Mosolapov, A.O. New numerical method for solving time-harmonic Maxwell equations with strong singularity. *J. Comput. Phys.* **2012**, *231*, 2438–2448, doi:10.1016/j.jcp.2011.11.031.
24. Rukavishnikov, V.A.; Mosolapov, A.O. Weighted edge finite element method for Maxwell's equations with strong singularity. *Dokl. Math.* **2013**, *87*, 156–159, doi:10.1134/S1064562413020105.
25. Rukavishnikov, V.A.; Rukavishnikov, A.V. Weighted finite element method for the Stokes problem with corner singularity. *J. Comput. Appl. Math.* **2018**, *341*, 144–156, doi:10.1016/j.cam.2018.04.014.
26. Rukavishnikov, V.A.; Rukavishnikov, A.V. New Numerical Method for the Rotation form of the Oseen Problem with Corner Singularity. *Symmetry* **2019**, *11*, 54, doi:10.3390/sym11010054.
27. Rukavishnikova, E.I. Numerical Method for the First Boundary Value Problem with Degeneration. In Proceedings of the International Conference "Computational mathematics, Differential Equations, Information Technologies", Lake Baikal, 22–24 August 2009; East Siberia State University of Technology and Management: Ulan-Ude, Russia, 2009; pp. 295–301.
28. Rukavishnikov, V.A.; Rukavishnikova, E.I. On the isomorphic mapping of weighted spaces by an elliptic operator with degeneration on the domain boundary. *Differ. Equ.* **2014**, *50*, 345–351, doi:10.1134/S0012266114030082.
29. Rukavishnikova, E.I. The automation tracing of mesh with condensed to the boundary of domain. *Inf. Sci. Control Syst.* **2011**, *30*, 57–64.
30. Liseikin, V.D. *Grid Generation Methods*; Springer: Berlin, Germany, 1999; 362p.
31. Apel, T.; Sandig, A.M.; Whiteman, J.R. Graded mesh refinement and error estimates for finite element solutions of elliptic boundary value problems in non-smooth domains. *Math. Methods Appl. Sci.* **1996**, *19*, 63–85.
32. Soghrati, S.; Xiao, F.; Nagarajan, A. A conforming to interface structured adaptive mesh refinement technique for modeling fracture problems. *Comput. Mech.* **2017**, *59*, 667–684, doi:10.1007/s00466-016-1366-z.
33. Rukavishnikov, V.A.; Rukavishnikova, H.I. Dirichlet Problem with Degeneration of the Input Data on the Boundary of the Domain. *Differ. Equ.* **2016**, *52*, 681–685, doi:10.1134/S0012266116050141.

© 2019 by the authors. Licensee MDPI, Basel, Switzerland. This article is an open access article distributed under the terms and conditions of the Creative Commons Attribution (CC BY) license (http://creativecommons.org/licenses/by/4.0/).

15. Rukavishnikov, V.A. On the existence and uniqueness of an R_ν-generalized solution of a boundary value problem with uncoordinated degeneration of the input data. *Dokl. Math.* **2014**, *90*, 562–564, doi:10.1134/S1064562414060155.
16. Rukavishnikov, V.A. The weight estimation of the speed of difference scheme convergence. *Dokl. Akad. Nauk SSSR* **1986**, *288*, 1058–1062.
17. Rukavishnikov, V.A.; Rukavishnikova, E.I. Finite-Element Method for the 1St Boundary-Value Problem with the Coordinated Degeneration of the Initial Data. *Dokl. Akad. Nauk* **1994**, *338*, 731–733.
18. Rukavishnikov, V.A. Methods of numerical analysis for boundary value problem with strong singularity. *Rus. J. Numer. Anal. Math. Model.* **2009**, *24*, 565–590, doi:10.1016/j.cam.2010.01.020.
19. Rukavishnikov, V.A.; Rukavishnikova, H.I. The finite element method for a boundary value problem with strong singularity. *J. Comput. Appl. Math.* **2010**, *234*, 2870–2882, doi:10.1515/RJNAMM.2009.035.
20. Rukavishnikov, V.A.; Rukavishnikova, H.I. On the error estimation of the finite element method for the boundary value problems with singularity in the Lebesgue weighted space. *Numer. Funct. Anal. Opt.* **2013**, *34*, 1328–1347, doi:10.1080/01630563.2013.809582.
21. Rukavishnikov, V.A.; Nikolaev, S.G. Weighted finite element method for an elasticity problem with singularity. *Dokl. Math.* **2013**, *88*, 705–709, doi:10.1134/S1064562413060215.
22. Rukavishnikov, V.A.; Rukavishnikova, H.I. *Weighted Finite-Element Method for Elasticity Problems with Singularity*, p razvan ed.; Finite Element Method. Simulation, Numerical Analysis and Solution Techniques, IntechOpen Limited: London, UK, 2018; pp. 295–311, doi:10.5772/intechopen.72733.
23. Rukavishnikov, V.A.; Mosolapov, A.O. New numerical method for solving time-harmonic Maxwell equations with strong singularity. *J. Comput. Phys.* **2012**, *231*, 2438–2448, doi:10.1016/j.jcp.2011.11.031.
24. Rukavishnikov, V.A.; Mosolapov, A.O. Weighted edge finite element method for Maxwell's equations with strong singularity. *Dokl. Math.* **2013**, *87*, 156–159, doi:10.1134/S1064562413020105.
25. Rukavishnikov, V.A.; Rukavishnikov, A.V. Weighted finite element method for the Stokes problem with corner singularity. *J. Comput. Appl. Math.* **2018**, *341*, 144–156, doi:10.1016/j.cam.2018.04.014.
26. Rukavishnikov, V.A.; Rukavishnikov, A.V. New Numerical Method for the Rotation form of the Oseen Problem with Corner Singularity. *Symmetry* **2019**, *11*, 54, doi:10.3390/sym11010054.
27. Rukavishnikova, E.I. Numerical Method for the First Boundary Value Problem with Degeneration. In Proceedings of the International Conference "Computational mathematics, Differential Equations, Information Technologies", Lake Baikal, 22–24 August 2009; East Siberia State University of Technology and Management: Ulan-Ude, Russia, 2009; pp. 295–301.
28. Rukavishnikov, V.A.; Rukavishnikova, E.I. On the isomorphic mapping of weighted spaces by an elliptic operator with degeneration on the domain boundary. *Differ. Equ.* **2014**, *50*, 345–351, doi:10.1134/S0012266114030082.
29. Rukavishnikova, E.I. The automation tracing of mesh with condensed to the boundary of domain. *Inf. Sci. Control Syst.* **2011**, *30*, 57–64.
30. Liseikin, V.D. *Grid Generation Methods*; Springer: Berlin, Germany, 1999; 362p.
31. Apel, T.; Sandig, A.M.; Whiteman, J.R. Graded mesh refinement and error estimates for finite element solutions of elliptic boundary value problems in non-smooth domains. *Math. Methods Appl. Sci.* **1996**, *19*, 63–85.
32. Soghrati, S.; Xiao, F.; Nagarajan, A. A conforming to interface structured adaptive mesh refinement technique for modeling fracture problems. *Comput. Mech.* **2017**, *59*, 667–684, doi:10.1007/s00466-016-1366-z.
33. Rukavishnikov, V.A.; Rukavishnikova, H.I. Dirichlet Problem with Degeneration of the Input Data on the Boundary of the Domain. *Differ. Equ.* **2016**, *52*, 681–685, doi:10.1134/S0012266116050141.

© 2019 by the authors. Licensee MDPI, Basel, Switzerland. This article is an open access article distributed under the terms and conditions of the Creative Commons Attribution (CC BY) license (http://creativecommons.org/licenses/by/4.0/).

Article

Finite Element Study on the Wear Performance of Movable Jaw Plates of Jaw Crushers after a Symmetrical Laser Cladding Path

Yuhui Chen [1,*], Guoshuai Zhang [1], Ruolin Zhang [1], Timothy Gupta [2,*] and Ahmed Katayama [3]

1. School of Energy and Power Engineering, Zhengzhou University of Light Industry, Zhengzhou 450000, Henan, China; 331821010411@zzuli.edu.cn (G.Z.); 331921020487@zzuli.edu.cn (R.Z.)
2. Queen Mary Research Institute of London, London E1 4NS, UK
3. School of Electronic Engineering and Mechanical, New York University of Technology, New York, NY 10027, USA; nyutkatayama@nyut.us
* Correspondence: chenyh@zzuli.edu.cn (Y.C.); mutimothy@qmril.gb (T.G.); Tel.: +1-006-662-8264 (T.G.)

Received: 19 April 2020; Accepted: 2 July 2020; Published: 7 July 2020

Abstract: At present, research on the influence of friction heat on the wear resistance of laser cladding layers is still lacking, and there is even less research on the temperature of laser cladding layers under different loads by a finite element program generator (FEPG). After a symmetrical laser cladding path, the wear performance of the moving jaw will change. The study of the temperature change of the moving jaw material in friction provides a theoretical basis for the surface modification of the moving jaw. The model of the column ring is built in a finite element program generator (FEPG). When the inner part of the column is WDB620 (material inside the cylinder) and the outer part is ceramic powder (moving jaw surface material), the relationship between the temperature and time of the contact surface is analyzed under the load between 100 and 600 N. At the same time, the stable temperature, wear amount, effective hardening layer thickness, strain thickness, and iron oxide content corresponding to different loads in a finite element program generator (FEPG) were analyzed. The results showed that when the load is 300 N, the temperature error between the finite element program generator (FEPG) and the movable jaw material is the largest, and the relative error is 4.3%. When the load increases, the stable temperature of the moving jaw plate increases after the symmetrical laser cladding path, and the wear amount first decreases and then increases. The minimum wear amount appears at a load of 400 N and a temperature of 340 °C; the strain thickness of the sample material increases gradually, and the effective hardening layer thickness increases. However, when the load reaches 400 N, the thickness of the effective hardening layer changes little; the content of Fe decreases gradually, and the content of FeO and Fe_2O_3 increases. The increase of the moving jaw increases in turn the temperature of the laser cladding layer of the test jaw material, which intensifies the oxidation reaction of the ceramic powder of the laser cladding layer.

Keywords: jaw crusher; symmetrical laser cladding path; FEPG; wear

1. Introduction

The mobile jaw plate of a jaw crusher is under severe impact in humid and high temperature environments. In the world, there are approximately 200–300 thousand movable jaw plates damaged by wear every year, and the consumption of steel is approximately 60–72 thousand tons [1]. Each year, this directly causes a loss of more than one billion dollars [2]. Therefore, mining enterprises pay a huge price for it. In order to strengthen its wear resistance, the surface of the moveable jaw plate is usually treated by laser cladding and other methods [3]. Laser cladding technology is an advanced surface modification technology, which involves many disciplines, such as metal materials, metallurgy, chemistry, and so on. It is found that the tensile strength and hardness of the coating are always higher than that of the substrate, and the performance is better than that of the substrate [4–7], no matter how the laser power is set in a certain range. However, the analysis of the effect of friction heat on the wear resistance of laser cladding materials remains to be discussed. Therefore, the analysis of the wear behavior and material structure of the moving jaw plate after the symmetrical laser cladding path under the action of reciprocating impact and friction has a positive significance for the study of the influence of temperature on the wear resistance of the moving jaw plate and the surface modification of the moving jaw plate.

For the study of wear of the jaw crusher and other mining crushers after laser cladding, the existing research methods are mainly theoretical calculation, test, and other methods [8–10]. Karan et al. analyzed the main wear area of the acceleration plate of the vertical impact crusher, changed the structural parameters and production parameters of the rotor, respectively, and explored the wear characteristics of the acceleration plate [11]. Drzymała et al. analyzed the mechanical characteristics and particle movement characteristics of the hammer head of the vertical shaft impact crusher through theoretical calculation and determined that impact wear is the main form of hammer head wear [12]. Limanskiy et al. analyzed the movement of the jaw plate of the jaw crusher through experiments and analyzed the wear on the jaw plate of the jaw crusher based on the microwear mechanism and failure of the jaw plate surface, so as to learn the reason for jaw plate wear when materials are broken [13]. Amanov et al. analyzed the change trends of the grain area, and hardness and wear of the chain wheel of the mine conveyor after treatment at different temperatures and times and finally obtained the best treatment temperature and time [14]. Baek et al. analyzed the kinematic characteristics of the materials in the cone crusher through theoretical calculation, solved the crushing process, and obtained the compression ratio, particle size distribution coefficient, and crushing pressure in each region of the crushing chamber [15]. Abuhasel et al. calculated and analyzed the wear of the impact crusher plate hammer after different laser cladding processes and determined the spot diameter and powder spreading speed of laser cladding in the early stage with the minimum wear amount [16]. Anticoi et al. conducted an experimental analysis on the temperature field of the friction stir welding process on the roller surface of the roller crusher and found that the friction coefficient decreased with the increase of the temperature, and the welding temperature increased with the increase of the concave angle of the shoulder [17]. Pei et al. analyzed the influence of particle size on the liner wear of the semi-autogenous mill via a theoretical calculation method. The results showed that the wear of the lining plate is especially significant when the particle size increases, the wear of the lining plate increases with the increase of particle size, and the kinetic energy obtained by large particle materials is far greater than that of small particle materials [18].

The wear amount calculated by theory is quite different from the actual result, and most of the conditions calculated by theory are ideal. In the physical prototype, the wear performance test of the components of the crusher is carried out, and the analysis of the wear characteristics of the components has great data disturbance, so it is difficult to extract the relevant data accurately and collect the data. Therefore, in this study, a finite element model of the cylinder-ring was built in the finite element program generator (FEPG). The temperature of laser cladding ceramic powder (surface material of the movable jaw plate) obtained by FEPG under different loads was verified and compared with the temperature obtained by the test. At the same time, the stable temperature, wear amount, effective hardening layer thickness, strain thickness, and iron oxide content corresponding to different loads were analyzed. The research results can be used to solve the problems of failure of the jaw crusher caused by high friction temperature and low service life of the moving jaw plate under complex working conditions. At the same time, they can provide a theoretical basis for analyzing the wear performance of the moving jaw plate material after symmetrical laser cladding path and the temperature change of the moving jaw plate in friction, which is of great significance to the surface modification of the moving jaw plate.

In addition, through the existing literature, it is found that in the cylinder disk test, along the direction of sliding friction, plastic deformation occurs on the surface of cylinder coating, resulting in different thicknesses of the work hardening layer [19–21]. However, analysis of the effect of friction heat on the wear resistance of laser cladding materials is still lacking.

2. Materials and Methods

In this study, WDB620 (Figure 1) was selected as the material of the moving jaw plate, and then the ceramic powder was laser-cladded on the surface of the moving jaw plate. The contents of various chemical components of ceramic powder are shown in Table 1 [22–25].

Figure 1. WDB620 material.

Table 1. Chemical composition.

Mn (%)	Rb (%)	C (%)	P (%)	V (%)	B (%)	Re (%)	Fe (%)
2.2	1.3	22.5	0.3	1.2	1.7	3.8	26.4

The cylinder sample after cutting is installed on the friction electronic tester and closely contacts with the ring. The material of the ring is 40Cr, the loads of the cylinder sample are 100, 200, 300, 400, 500, and 600 N, and the rotation speed of the ring is 2000 r/min. The test scheme was carried out according to Table 2.

Table 2. Test plan.

Serial Number	Load X_1/N	Time X_2/s	Speed X_3/s
1	100	550	2000
2	200	550	2000
3	300	550	2000
4	400	550	2000
5	500	550	2000
6	600	550	2000

Through the thermocouple temperature sensor used to collect the temperature of the cylinder sample (Figure 2), the collected temperature of each point is counted.

The model of the cylinder-ring was built in UG (Figure 3). (UG is an interactive computer-aided design and computer-aided manufacturing software, which has powerful functions and can realize the construction of various complex entities and models. It is mainly used in the product development fields of mechanical product modeling design, structural design, part assembly design, mold design, numerical control programming, design analysis, etc.) The radius difference between the two concentric circles is 230 mm [26–30], and it is added to the creator in FEPG in the form of x_t (Figure 4). The material of the setting ring in FEPG is 40Cr, the inner part of the cylinder is WDB620 (movable jaw plate material), and the outer part is laser cladding ceramic powder. The specific physical parameter settings of the cylinder-ring are shown in Table 3 [31–34].

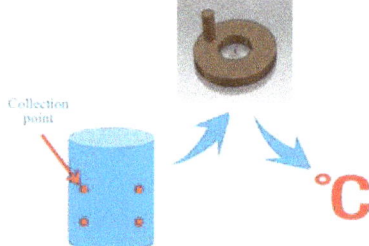

Figure 2. Thermocouple temperature sensor.

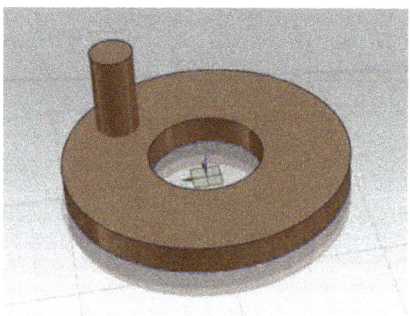

Figure 3. Cylinder-ring model in UG.

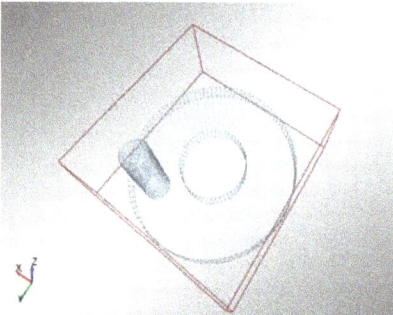

Figure 4. Cylinder-ring model in a finite element program generator (FEPG).

Table 3. Parameter setting in FEPG.

Material	40Cr	WDB620	Ceramic Powder
Shear modulus (Pa)	2.06×10^{12}	2.26×10^{11}	4.16×10^{11}
Heat transfer coefficient (W/(m·k))	35	60	72
Density (kg/m^3)	7220	8230	7850

3. Results and Discussion

3.1. FEPG and Test Results

The finite element study [35–37] was carried out under the loads of 100, 200, 300, 400, 500, and 600 N, respectively, and the rotating speed of the ring was 2000 r/min. The maximum surface temperature and radial diffusion size of the ring under different loads were obtained (Figure 5).

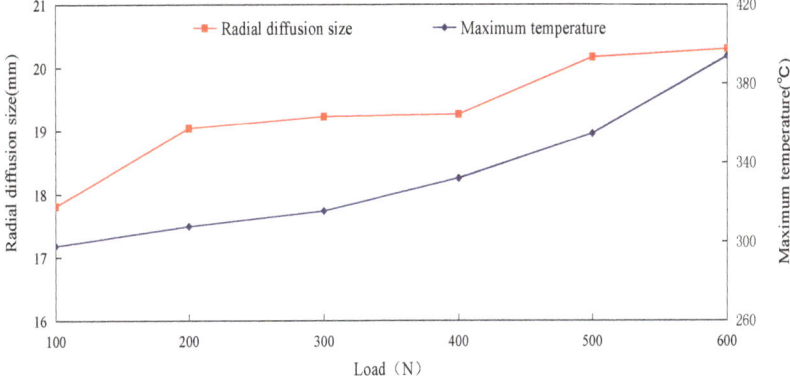

Figure 5. Maximum value and radial diffusion size of ring surface temperature.

The maximum temperature and radial diffusion size in Figure 5 are analyzed. When the cylinder loads are 100, 200, 300, 400, 500, and 600 N, the maximum corresponding temperature is 396.8 °C and the radial diffusion size is 20.3 mm. With the increase of the load on the cylinder, the maximum temperature and the radial diffusion size of the ring surface are increasing.

When the cylinder loads are 100, 200, 300, 400, 500, and 600 N, the temperature change rule of the cylinder-ring contact surface is studied (Figure 6).

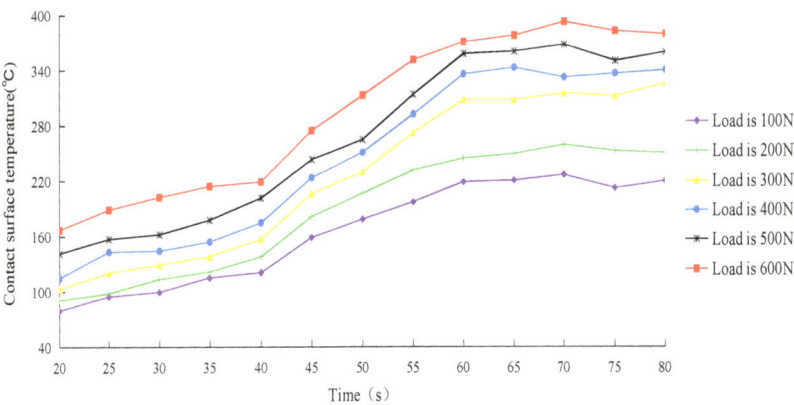

Figure 6. Temperature of cylinder-ring contact surface.

The temperature of the contact surface in Figure 6 is analyzed. At the same time, the temperature of the contact surface of the cylinder increases with the increase of the load on the cylinder. However, the change trend of contact surface temperature under different loads is basically the same. After the time reaches about 60 s, the temperature does not rise significantly. When the cylinder loads are 100,

200, 300, 400, 500, and 600 N, the contact surface temperature is stable at 222, 251, 326, 341, 360, and 378 °C, respectively.

Through the test, the mean value of the temperature at the acquisition point under the sample loads of 100, 200, 300, 400, 500, and 600 N is obtained. The test result data are processed by high-order function regression through SPSS software (12.0, SPSS Inc., Chicago, IL, USA), and the regression function of temperature relative to time and load is as follows:

$$L = 600.421 - 3.252X_1 - 8.002X_2 + 1.763X_1X_2 - 0.287X_1^2 - 0.891X_2^2 \qquad (1)$$

The temperature of the contact surface under different time and load is shown in Figure 7.

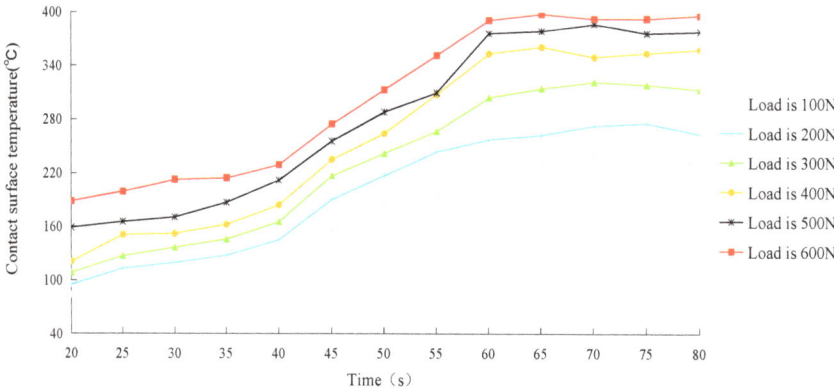

Figure 7. Temperature obtained at different times.

From Figure 7, when the load of the sample increases from 100 to 600 N, the temperature corresponding to the fixed time increases. When the loads of the sample are 100, 200, 300, 400, 500, and 600 N, the temperature of test and FEPG changes with time (Figure 8).

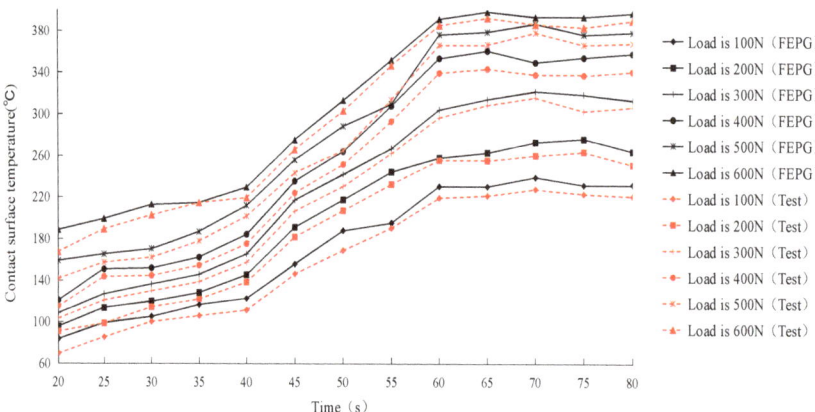

Figure 8. Comparison between test and FEPG.

From Figure 8, when the sample loads are 100, 200, 300, 400, 500, and 600 N, the temperature obtained by FEPG at the same time point is slightly higher than the test result. The reason for the result is that there are many environmental effects, such as heat and air transfer loss, in the actual test.

However, when the loads of FEPG are 100, 200, 300, 400, 500, and 600 N, the trend of temperature change is the same, and when the load is 400 N and the time is 80 s, the temperature error is the largest, and the relative error is only 4.3%.

3.2. Effect of Temperature on the Performance of the Moving Jaw

According to the previous test and FEPG study, after a certain period of time (this study is about 65 s), the temperature of the moving jaw material sample tends to be stable, so the performance of the stabilized moving jaw material is analyzed. The steady temperature and wear of the cylinder under different loads are analyzed (Figure 9).

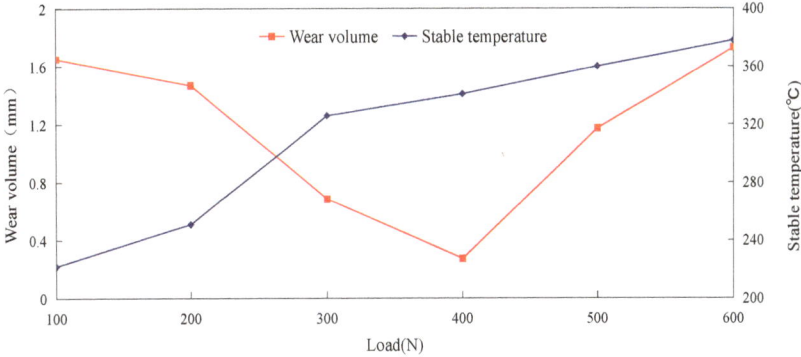

Figure 9. Change trend of temperature and wear.

With the increase of load, the stable temperature rises gradually, and the wear amount rises after an initial decrease. When the load is at 400 N and the temperature is around 340 °C, the value is the smallest. This is because the laser cladding layer on the surface of the sample material (moving jaw material) is partially melted under high heat, which reduces the wear. However, the temperature and load continue to increase, which not only destroys the protective layer, but also causes part of the remaining protective layer to bond and tear due to the rotation of the ring.

Through the electron microscope, the effective hardening layer thickness and strain thickness of the moving jaw specimen under different loads were analyzed, and the results were compared (Figure 10).

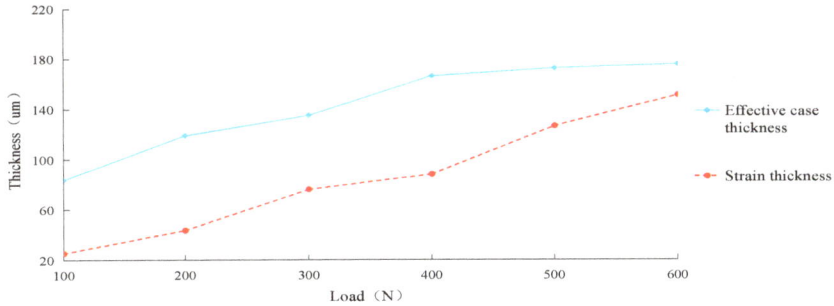

Figure 10. Effective hardening layer thickness and strain thickness

Through the analysis of the effective hardening layer thickness and strain thickness under no load, it is known that with the increase of load, the strain thickness of the sample material increases gradually, and the effective hardening layer thickness increases significantly, but when the load reaches

400 N, the effective hardening layer thickness changes little. The results show that when the load is about 400 N, the wear resistance of the laser cladding material is relatively good.

The surface of the sample was analyzed by X-ray photoelectron spectroscopy (XPS), and the content changes of Fe, O, Mn, C, V, and Re were counted (Figure 11).

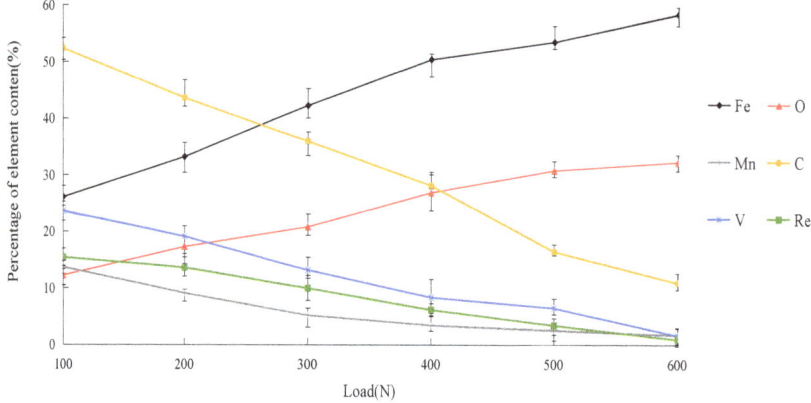

Figure 11. Content change of Fe, O, Ti, Cr, V, and Nb.

It can be seen from Figure 11 that there are mainly Fe, O, Mn, V, and other elements on the surface of the wear mark of the material, and the mass fraction of Fe and O elements accounts for the largest proportion. With the increase of the load, the mass fraction of Fe element increases from 26.08% to 58.42%, and the mass fraction of O element increases from 12.28% to 32.37%. With the increase of the load on the sample, the content of Fe and O increases gradually. In the process of friction, the increase of the load on the cylinder causes the temperature of the friction contact surface of the cylinder to rise, and the constant formation of iron oxide, which leads to the oxidation and wear of the deposition layer of the cylinder.

Analyzing the content of different types of iron oxide in the laser cladding layer of the moving jaw plate, A represents Fe, B represents FeO, and C represents Fe_2O_3 (Figure 12).

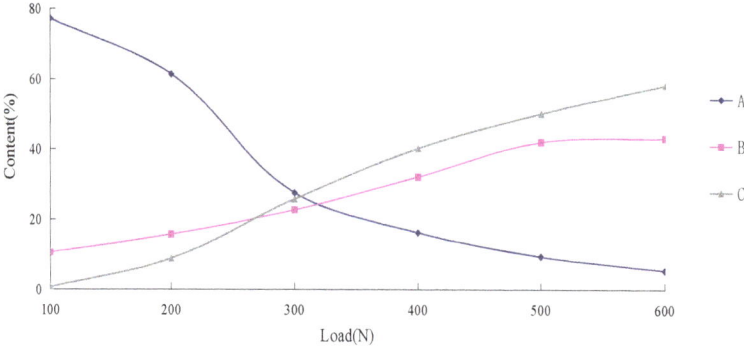

Figure 12. Content of Fe, FeO, and Fe_2O_3.

The contents of Fe, FeO, and Fe_2O_3 were studied. It was found that when the load was 100 N, the oxide on the worn surface was mainly Fe, and then there is not only FeO, but also Fe_2O_3 when the load is 200 and 300 N. With the increase of load from 400 to 600 N, the content of FeO and Fe_2O_3 on the worn surface of the coating continued to increase. In the above analysis, it can be found that with the increase of friction temperature, the high-temperature transformation process of wear on the surface is Fe → FeO → Fe$_2$O$_3$. The main reason is that in the process of friction and wear, the micro convex particles on the contact surface easily produce a flash point temperature (higher than 1500°) under the conditions of high load and high speed, and the changes of temperature and load affect the change of oxide type [38–40]. The simulation experiment is in an ideal working environment where there are no micro convex particles on the surface of the friction pair and there is no flash point high temperature. In the experiment of temperature measurement, the temperature changes dynamically, and the temperature of the contact surface is the highest. The thermocouple only measures the surface temperature, so it cannot measure the flash point temperature. Thus, the main reason for the existence of the iron oxide is that the heat generated by friction is accumulating in the contact surface, which leads to the plastic deformation of the material surface caused by oxidation and softening, and the appearance of the friction oxide layer. The results show that the increase of the sample load increases the temperature of the laser cladding layer of the sample material (movable jaw plate material), which intensifies the oxidation reaction of the ceramic powder of the laser cladding layer.

4. Conclusions

Based on the theory of tribology and heat transfer, focusing on the influence of temperature on the properties of laser cladding ceramic powder (surface material of the movable jaw plate), the coupling analysis of cylinder disc wear is completed using the finite element software FEPG, and a temperature measurement test bench which is built on the wear test machine for test analysis and verification. Finally, the wear amount, effective hardening layer thickness, and strain thickness of the moving jaw after symmetrical laser cladding path are analyzed by FEPG. It is concluded that in the wear process of the moving jaw material after the symmetrical laser cladding path, the maximum temperature of the moving jaw material is related to its load.

It is found that when the load is 100, 200, 300, 400, 500, and 600 N and the time is about 65 s, the contact surface temperature is stable at 222, 251, 326, 341, 360, and 378 °C, respectively. The results show that the temperature of FEPG at the same time point is slightly higher than that of the test when the loads of the sample are 100, 200, 300, 400, 500, and 600 N, but the trend of temperature change is the same. When the load is 300 N, the temperature error between the test and FEPG is the largest, and the relative error is only 4.3%. When the load increases, the stable temperature of the laser-cladded moving jaw material increases, and the wear amount first decreases and then increases. The minimum wear amount appears at the load of 400 N, and the temperature is about 320 °C. With the increase of load, the thickness of strain increases gradually, and the thickness of the effective hardening layer increases obviously, but when the load reaches 400 N, the thickness of the effective hardening layer changes little. The increase of the sample load increases in turn the temperature of the laser cladding layer of the sample material, which intensifies the oxidation reaction of the ceramic powder of the laser cladding layer. The research results provide a new idea for the analysis of the wear behavior of the moving jaw plate after laser cladding under the action of reciprocating impact and friction and provide a theoretical basis for the analysis of the change of the wear resistance of the moving jaw plate caused by the temperature and the surface modification of the moving jaw plate.

Author Contributions: Conceptualization, Y.C. and T.G.; data curation, Y.C., G.Z., R.Z., T.G. and A.K.; formal analysis, Y.C., G.Z., R.Z., T.G. and A.K.; funding acquisition, Y.C.; investigation, Y.C.; resources, Y.C.; software, Y.C., G.Z., R.Z. and T.G.; supervision, Y.C. and T.G.; visualization, Y.C. and G.Z.; writing—original draft preparation, Y.C., G.Z., R.Z., T.G. and A.K.; writing—review and editing, Y.C., G.Z., R.Z. and T.G. All authors have read and agreed to the published version of the manuscript.

Funding: This work is supported by the National Natural Science Foundation of China Youth Foundation (no.51705474).

Acknowledgments: The authors thank the anonymous editor for the editing assistance. Lastly, the authors would like to thank the anonymous reviewers for their valuable comments and suggestions on an earlier version of our manuscript.

Conflicts of Interest: The authors declare that they have no conflict of interest.

References

1. Jiang, J.; Xie, Y.; Qian, W.; Hall, R. Important factors affecting the gouging abrasion resistance of materials. *Miner. Eng.* **2018**, *128*, 238–246. [CrossRef]
2. Deepak, B.B.; Bahubalendruni, M.V. Numerical analysis for force distribution along the swing jaw plate of a single toggle jaw crusher. *World J. Eng.* **2017**, *14*, 255–260. [CrossRef]
3. Liu, R.; Shi, B.; Li, G.; Yu, H. Influence of operating conditions and crushing chamber on energy consumption of cone crusher. *Energies* **2018**, *11*, 1102. [CrossRef]
4. Li, Y.; Zhang, P.; Bai, P.; Su, K.; Su, H. TiBCN-Ceramic-Reinforced Ti-Based Coating by Laser Cladding: Analysis of Processing Conditions and Coating Properties. *Coatings* **2019**, *9*, 407. [CrossRef]
5. Gabriel, T.; Rommel, D.; Scherm, F.; Gorywoda, M.; Glatzel, U. Symmetrical laser cladding path of ultra-thin nickel-based superalloy sheets. *Materials* **2017**, *10*, 279. [CrossRef]
6. Yan, X.-L.; Dong, S.-Y.; Xu, B.-S.; Cao, Y. Progress and challenges of ultrasonic testing for stress in remanufacturing symmetrical laser cladding path coating. *Materials* **2018**, *11*, 293. [CrossRef]
7. Xu, Z.; Wang, Z.; Chen, J.; Qiao, Y.; Zhang, J.; Huang, Y. Effect of rare earth oxides on microstructure and corrosion behavior of laser-cladding coating on 316l stainless steel. *Coatings* **2019**, *9*, 636. [CrossRef]
8. Pang, X.; Wei, Q.; Zhou, J.; Ma, H. High-temperature tolerance in multi-scale cermet solar-selective absorbing coatings prepared by symmetrical laser cladding path. *Materials* **2018**, *11*, 1037. [CrossRef]
9. Lian, G.; Zhang, H.; Zhang, Y.; Tanaka, M.L.; Chen, C.; Jiang, J. Optimizing processing parameters for multi-track symmetrical laser cladding path utilizing multi-response grey relational analysis. *Coatings* **2019**, *9*, 356. [CrossRef]
10. Wang, S.; Liu, C. Real-time monitoring of chemical composition in nickel-based symmetrical laser cladding path layer by emission spectroscopy analysis. *Materials* **2019**, *12*, 2637. [CrossRef]
11. Karan, S.K.; Samadder, S.R. Reduction of spatial distribution of risk factors for transportation of contaminants released by coal mining activities. *J. Environ. Manag.* **2016**, *180*, 280–290. [CrossRef] [PubMed]
12. Drzymała, T.; Zegardło, B.; Tofilo, P. Properties of concrete containing recycled glass aggregates produced of exploded lighting materials. *Materials* **2020**, *13*, 226. [CrossRef] [PubMed]
13. Limanskiy, A.V.; Vasilyeva, M.A. Using of low-grade heat mine water as a renewable source of energy in coal-mining regions. *Ecol. Eng.* **2016**, *91*, 41–43. [CrossRef]
14. Amanov, A.; Ahn, B.; Lee, M.G.; Jeon, Y.; Pyun, Y. Friction and wear reduction of eccentric journal bearing made of sn-based babbitt for ore cone crusher. *Materials* **2016**, *9*, 950. [CrossRef]
15. Baek, J.; Choi, Y. Deep neural network for ore production and crusher utilization prediction of truck haulage system in underground mine. *Appl. Sci.* **2019**, *9*, 4180. [CrossRef]
16. Abuhasel, K.A. A comparative study of regression model and the adaptive neuro-fuzzy conjecture systems for predicting energy consumption for jaw crusher. *Appl. Sci.* **2019**, *9*, 3916. [CrossRef]
17. Anticoi, H.; Guasch, E.; Ahmad Hamid, S.; Oliva, J.; Alfonso, P.; Bascompta, M.; Sanmiquel, L.; Escobet, T.; Escobet, A.; Parcerisa, D.; et al. An improved high-pressure roll crusher model for tungsten and tantalum ores. *Minerals* **2018**, *8*, 483. [CrossRef]
18. Pei, Z.; Galindo-Torres, S.A.; Tang, H.; Jin, G.; Scheuermann, A.; Ling, L. An efficient discrete element lattice boltzmann model for simulation of particle-fluid, particle-particle interactions. *Comput. Fluids* **2017**, *147*, 63–71.
19. Ren, X.; Liu, Z. Influence of cutting parameters on work hardening behavior of surface layer during turning superalloy Inconel 718. *Int. J. Adv. Manuf. Technol.* **2016**, *86*, 2319–2327. [CrossRef]
20. Gao, P.; Wei, K.; Hanchen, Y.U.; Yang, J.; Wang, Z.; Zeng, X. Influence of layer thickness on microstructure and mechanical properties of selective laser melted Ti-5Al-2.5Sn alloy. *Acta Metall. Sin.* **2018**, *54*, 999–1009.
21. Nishida, Y.; Yokoyama, S. Mechanisms of temperature dependence of threshold voltage in high-k/metal gate transistors with different TiN thicknesses. *Int. J. Electron.* **2016**, *103*, 629–647. [CrossRef]
22. Cuomo, S.; Chareyre, B.; D'Arista, P.; Sala, M.D.; Cascini, L. Micromechanical modelling of rainsplash erosion in unsaturated soils by finite element method. *Catena* **2016**, *147*, 146–152. [CrossRef]
23. Kroupa, M.; Vonka, M.; Soos, M.; Kosek, J. Utilizing the finite element method for the modeling of viscosity in concentrated suspensions. *Langmuir* **2016**, *32*, 8451–8460. [CrossRef] [PubMed]

24. Fallahnezhad, K.; Oskouei, R.H.; Badnava, H.; Taylor, M. An adaptive finite element simulation of fretting wear damage at the head-neck taper junction of total hip replacement: The role of taper angle mismatch. *J. Mech. Behav. Biomed. Mater.* **2017**, *75*, 58–67. [CrossRef]
25. Shinde, T.; Dhokey, N.B. Influence of carbide density on surface roughness and quasi-stable wear behaviour of h13 die steel. *Surf. Eng.* **2017**, *33*, 1–9. [CrossRef]
26. Soltani, R.; Sohi, M.H.; Ansari, M.; Haghighi, A.; Ghasemi, H.M.; Haftlang, F. Evaluation of niobium carbide coatings produced on aisi l2 steel via thermo-reactive diffusion technique. *Vacuum* **2017**, *146*, 44–51. [CrossRef]
27. Liu, G.; Sun, W.C.; Lowinger, S.M.; Zhang, Z.H.; Ming, H.; Peng, J. Coupled flow network and finite element modeling of injection-induced crack propagation and coalescence in brittle rock. *Acta Geotech.* **2018**, *14*, 1–26.
28. Thomas, J.J.; Ii, J.J.V.; Craddock, P.R.; Bake, K.D.; Pomerantz, A.E. The neutron scattering length density of kerogen and coal as determined by ch 3 oh/cd 3 oh exchange. *Fuel* **2014**, *117*, 801–808. [CrossRef]
29. Indimath, S.; Shunmugasundaram, R.; Balamurugan, S.; Das, B.; Singh, R.; Dutta, M. Ultrasonic technique for online measurement of bulk density of stamp charge coal cakes in coke plants. *Fuel Process. Technol.* **2018**, *172*, 155–161. [CrossRef]
30. Wang, Q.; Zeng, X.; Chen, C.; Lian, G.; Huang, X. An integrated method for multi-objective optimization of multi-pass Fe50/TiC symmetrical laser cladding path on AISI 1045 steel based on grey relational analysis and principal component analysis. *Coatings* **2020**, *10*, 151. [CrossRef]
31. Yu, J.; Sun, W.; Huang, H.; Wang, W.; Wang, Y.; Hu, Y. Crack sensitivity control of nickel-based laser coating based on genetic algorithm and neural network. *Coatings* **2019**, *9*, 728. [CrossRef]
32. Wu, F.; Chen, T.; Wang, H.; Liu, D. Effect of Mo on microstructures and wear properties of in situ synthesized Ti(C,N)/Ni-based composite coatings by symmetrical laser cladding path. *Materials* **2017**, *10*, 1047. [CrossRef] [PubMed]
33. Li, C.; Liu, C.; Li, S.; Zhang, Z.; Zeng, M.; Wang, F.; Wang, J.; Guo, Y. Numerical simulation of thermal evolution and solidification behavior of symmetrical laser cladding path AlSiTiNi composite coatings. *Coatings* **2019**, *9*, 391. [CrossRef]
34. Li, Y.; Chen, T.; Liu, D. Path planning for symmetrical laser cladding path robot on artificial joint surface based on topology reconstruction. *Algorithms* **2020**, *13*, 93. [CrossRef]
35. Wang, Y.; Liang, Z.; Zhang, J.; Ning, Z.; Jin, H. Microstructure and antiwear property of symmetrical laser cladding path Ni–Co duplex coating on copper. *Materials* **2016**, *9*, 634. [CrossRef]
36. Yang, K.; Xie, H.; Sun, C.; Zhao, X.; Li, F. Influence of vanadium on the microstructure of IN718 alloy by symmetrical laser cladding path. *Materials* **2019**, *12*, 3839. [CrossRef]
37. Chen, J.; Zhou, Y.; Shi, C.; Mao, D. Microscopic analysis and electrochemical behavior of Fe-based coating produced by symmetrical laser cladding path. *Metals* **2017**, *7*, 435. [CrossRef]
38. Lu, H.; Zhang, P.; Ren, S.; Guo, J.; Li, X.; Dong, G.I. The preparation of polytrifluorochloroethylene (PCTFE) micro-particles and application on treating bearing steel surfaces to improve the lubrication effect for copper-graphite (Cu/C). *Appl. Surf. Sci.* **2018**, *427*, 1242–1247. [CrossRef]
39. Zori, M.H.; Soleimanigorgani, A. Ink-jet printing of micro-emulsion TiO2 nano-particles ink on the surface of glass. *Ournal Eur. Ceram. Soc.* **2012**, *32*, 4271–4277. [CrossRef]
40. Feigenbaum, E.; Raman, R.N.; Cross, D.A.; Carr, C.W.; Matthews, M.J. Laser-induced Hertzian fractures in silica initiated by metal micro-particles on the exit surface. *Opt. Express* **2016**, *24*, 10527–10536. [CrossRef]

© 2020 by the authors. Licensee MDPI, Basel, Switzerland. This article is an open access article distributed under the terms and conditions of the Creative Commons Attribution (CC BY) license (http://creativecommons.org/licenses/by/4.0/).

Article

Numerical Solution of Direct and Inverse Problems for Time-Dependent Volterra Integro-Differential Equation Using Finite Integration Method with Shifted Chebyshev Polynomials

Ratinan Boonklurb * , **Ampol Duangpan and Phansphitcha Gugaew**

Department of Mathematics and Computer Science, Faculty of Science, Chulalongkorn University, Bangkok 10330, Thailand; ty_math@hotmail.com (A.D.); poohdd28@hotmail.com (P.G.)
* Correspondence: ratinan.b@chula.ac.th

Received: 3 February 2020; Accepted: 5 March 2020; Published: 30 March 2020

Abstract: In this article, the direct and inverse problems for the one-dimensional time-dependent Volterra integro-differential equation involving two integration terms of the unknown function (i.e., with respect to time and space) are considered. In order to acquire accurate numerical results, we apply the finite integration method based on shifted Chebyshev polynomials (FIM-SCP) to handle the spatial variable. These shifted Chebyshev polynomials are symmetric (either with respect to the point $x = \frac{L}{2}$ or the vertical line $x = \frac{L}{2}$ depending on their degree) over $[0, L]$, and their zeros in the interval are distributed symmetrically. We use these zeros to construct the main tool of FIM-SCP: the Chebyshev integration matrix. The forward difference quotient is used to deal with the temporal variable. Then, we obtain efficient numerical algorithms for solving both the direct and inverse problems. However, the ill-posedness of the inverse problem causes instability in the solution and, so, the Tikhonov regularization method is utilized to stabilize the solution. Furthermore, several direct and inverse numerical experiments are illustrated. Evidently, our proposed algorithms for both the direct and inverse problems give a highly accurate result with low computational cost, due to the small number of iterations and discretization.

Keywords: finite integration method; shifted Chebyshev polynomial; direct and inverse problems; Volterra integro-differential equation; Tikhonov regularization method

MSC: 65R20; 65R32

1. Introduction

An integro-differential equation (IDE) is an equation which contains both derivatives and integrals of an unknown function. Several situations in the branches of science and engineering can be demonstrated by developing mathematical models which are often in the form of IDEs, such as in RLC circuit analysis, the activity of interacting inhibitory and excitatory neurons, the Wilson–Cowan model, and so on; see Reference [1] for more applications. In fact, many of these problems cannot be directly solved, because we may not know all necessary information or an incomplete system may be provided. This has led to the study of both direct and inverse problems for a certain type of one-dimensional IDE involving time, which is called the one-dimensional time-dependent Volterra IDE (TVIDE). Hence, in this study, we investigate the TVIDE of the following form

$$v_t(x,t) + \mathcal{L}v(x,t) = \int_0^t \kappa_1(x,\eta) v(x,\eta) d\eta + \int_0^x \kappa_2(\xi,t) v(\xi,t) d\xi + F(x,t), \qquad (1)$$

for all $(x,t) \in (0, L) \times (0, T]$, where x and t represent space and time variables, respectively; $\mathcal{L}$ is the spatial linear differential operator of order n; $\kappa_1(x,t)$ and $\kappa_2(x,t)$ are the given continuously integrable kernel functions; and $v(x,t)$ is an unknown function, which is to be determined subject to prescribed initial and boundary conditions. We remark that, if a forcing term $F(x,t)$ of (1) is given, then this problem has only one unknown $v(x,t) \in C^{n,1}([0,L] \times [0,T])$ to be solved, and it is called a direct problem. In contrast, if the forcing term $F(x,t)$ is missing, then this problem has two unknowns $F(x,t) \in C([0,L] \times [0,T])$ and $v(x,t) \in C^{n,1}([0,L] \times [0,T])$ to be solved, and it is called an inverse problem. However, for the inverse problem in this paper, we specifically define the forcing term $F(x,t) := \beta(t)f(x,t)$, where $\beta(t)$ is a missing source function to be retrieved and $f(x,t)$ is the given function. We note that (1) has both $\int_0^t \kappa_1(x,\eta)v(x,\eta)d\eta$ and $\int_0^x \kappa_2(\xi,t)v(\xi,t)d\xi$, while several studies in the literature have considered similar problems containing only one of these two terms.

The Volterra IDE containing only an integration term with respect to time arises in many applications, including the compression of poro-viscoelastic media, blow-up problems, analysis of space–time-dependent nuclear reactor dynamics, and so on; see Reference [2]. The existence, uniqueness, and asymptotic behavior of its solution have been discussed in Reference [3]. There are many authors who have studied the numerical solution of this type of problem by using techniques such as the finite element method [2], finite difference method (FDM) [4], collocation methods in polynomial spline [5], the implicit Runge–Kutta–Nyström method [6], the Legendre collocation method [7], and so on.

On the other hand, the Volterra IDE containing only an integration term with respect to space has also been studied in various areas, such as for the one-dimensional viscoelastic problem and one-dimensional heat flow in materials with memory [8], modeling heat/mass diffusion processes, biological species coexisting together with increasing and decreasing rates of growth, electromagnetism, and ocean circulation, among others [9]. Moreover, the existence and uniqueness for this type of Volterra IDE were shown in Reference [8]. Consequently, abundant numerical methods have appeared for finding solutions to this type of Volterra IDE using, for example, spline collocation method [10], collocation method with implicit Runge–Kutta method [11], decomposition method [12], and so on.

However, our problem deals with a Volterra IDE involving both temporal and spatial integrations. There have been no results in the literature regarding the existence and uniqueness of solutions to this type of problem. In this paper, we concentrate on providing a decent numerical procedure to find approximate solutions for both the direct and inverse problems of the proposed TVIDE (1).

Generally, it is well-known that the classification of problems involving differential equations was defined by Hadamard [13] in 1902. Mathematical problems involving differential equations are well-posed if the following conditions hold: existence, uniqueness, and stability. Otherwise, the problem is called ill-posed; this normally occurs in the inverse problem. Even though the initial and boundary conditions are prescribed, it is not sufficient to guarantee that our inverse problem (1) has unique solutions $\beta(t)$ and $v(x,t)$. Hence, additional conditions (e.g., the observation or measurement of data) need to be involved. In practice, there are many kinds of additional conditions; for example, a fixed point of the system, an average time of the system, or an integral of the system. After the additional conditions has been added as an auxiliary condition in our inverse problem (1), we can obtain the existence and uniqueness of $\beta(t)$ and $v(x,t)$. However, the additional condition may contains measurement or observation errors, which may cause the instability in the solutions; namely, a small perturbation in the input data can produce a considerable error, especially for $\beta(t)$. Thus, some regularization techniques are required to overcome the ill-posedness and stabilize the solution.

There exist many schemes which are generally used to solve both direct and inverse problems of Volterra IDEs, such as the above-mentioned methods. However, those methods utilize the process of approximating differentiation. It is well-known that numerical differentiation is very sensitive to rounding errors, as its manipulation task involves division by a small step-size. On the other hand, the process of numerical integration involves multiplication by a small step-size and, so, it is very insensitive to rounding errors. In recent years, the finite integration method (FIM) has been developed to find approximate solutions of linear boundary value problems for partial differential equations (PDEs). The concept of FIM is to transform a given PDE into an equivalent integral equation, following which a numerical integration method, such as the trapezoid, Simpson, or Newton–Cotes methods (see References [14–16]), are applied. In 2018, Boonklurb et al. [17] modified the traditional FIM by using Chebyshev polynomials to solve one- and two-dimensional linear PDEs and obtained a more accurate result compared to the traditional FIMs and FDM. However, their technique [17] has not yet been utilized to overcome the direct and inverse problems of TVIDE, which are the major focuses of this work.

In this paper, we formulate numerical algorithms for solving the direct and inverse problems of TVIDE (1). We manipulate the idea of FIM in Reference [17] by using shifted Chebyshev polynomials, which we call the FIM with shifted Chebyshev polynomials (FIM-SCP), to deal with the spatial variable and use the forward difference quotient to estimate the time derivative. We further apply the Tikhonov regularization method to stabilize our ill-posed problem (1). The rest of the paper is organized as follows. In Section 2, the definition and some basic properties concerning the shifted Chebyshev polynomial are given to construct the shifted Chebyshev integration matrices. The Tikhonov regularization method is also presented in Section 2. In Section 3, we use the FIM-SCP and the forward difference quotient to devise efficient numerical algorithms to find approximate solutions to the direct and inverse problems of (1). Then, we implement our proposed algorithms through several examples, in order to demonstrate their efficiency compared with their analytical solutions. Furthermore, we also display the time convergence rate and CPU time (s) in Section 4. Finally, the conclusion and some directions for future work are given in Section 5.

2. Preliminaries

In this section, we introduce some necessary tools for solving the direct and inverse problems of TVIDE (1): the FIM-SCP and the Tikhonov regularization method.

2.1. Shifted Chebyshev Integration Matrices

We first introduce the definition and some basic properties of shifted Chebyshev polynomials [18], which are used to establish the first- and higher-order shifted Chebyshev integration matrices based on the idea of constructing integration matrices in Reference [17]. However, we slightly modify this idea by instead using a shifted Chebyshev expansion suitable for solving our problem (1) without domain transformation. We give the definition and properties as follows.

Definition 1. *The shifted Chebyshev polynomial of degree $n \geq 0$ is defined by*

$$S_n(x) = \cos\left(n \arccos\left(\frac{2x}{L} - 1\right)\right) \text{ for } x \in [0, L].$$

Note that this shifted Chebyshev polynomial is symmetric, either with respect to the point $x = \frac{L}{2}$ or the vertical line $x = \frac{L}{2}$ over $[0, L]$, depending on its degree. Next, we provide some important properties of the shifted Chebyshev polynomial, which we use to constructing the shifted Chebyshev integration matrix, as follows.

Lemma 1. (i) *For $n \in \mathbb{N}$, the zeros of $S_n(x)$ are symmetrically distributed over $[0, L]$ and given by*

$$x_k = \frac{L}{2}\left(\cos\left(\frac{2k-1}{2n}\pi\right) + 1\right), \, k \in \{1, 2, 3, ..., n\}. \tag{2}$$

(ii) *For $r \in \mathbb{N}$, the rth-order derivatives of $S_n(x)$ at the endpoint $b \in \{0, L\}$ are*

$$\left.\frac{d^r}{dx^r} S_n(x)\right|_{x=b} = \prod_{k=0}^{r-1}\left(\frac{n^2 - k^2}{2k+1}\right)\left(\frac{2b}{L} - 1\right)^{n+r}. \tag{3}$$

(iii) For $x \in [0, L]$, the single-layer integrations of shifted Chebyshev polynomial $S_n(x)$ are

$$\bar{S}_0(x) = \int_0^x S_0(\xi)\,d\xi = x,$$

$$\bar{S}_1(x) = \int_0^x S_1(\xi)\,d\xi = \frac{x^2}{L} - x,$$

$$\bar{S}_n(x) = \int_0^x S_n(\xi)\,d\xi = \frac{L}{4}\left(\frac{S_{n+1}(x)}{n+1} - \frac{S_{n-1}(x)}{n-1} - \frac{2(-1)^n}{n^2-1}\right),\ n \in \{2,3,4,\ldots\}.$$

(iv) Let $\{x_k\}_{k=1}^n$ be a set of zeros of $S_n(x)$, the shifted Chebyshev matrix $\mathbf{S}$ is defined by

$$\mathbf{S} = \begin{bmatrix} S_0(x_1) & S_1(x_1) & \cdots & S_{n-1}(x_1) \\ S_0(x_2) & S_1(x_2) & \cdots & S_{n-1}(x_2) \\ \vdots & \vdots & \ddots & \vdots \\ S_0(x_n) & S_1(x_n) & \cdots & S_{n-1}(x_n) \end{bmatrix}.$$

Then, it has the multiplicative inverse $\mathbf{S}^{-1} = \frac{1}{n}\mathrm{diag}(1,2,2,\ldots,2)\mathbf{S}^\top$.

Next, we use the above definition and properties of shifted Chebyshev polynomials to construct the shifted Chebyshev integration matrices. First, let N be a positive integer and L be a positive real number. Define an approximate solution $u(x)$ of a certain differential equation by a linear combination of shifted Chebyshev polynomials $S_n(x)$; that is,

$$u(x) = \sum_{n=0}^{N-1} c_n S_n(x) \text{ for } x \in [0, L]. \tag{4}$$

Let x_k for $k \in \{1,2,3,\ldots,N\}$ be the interpolated points which are meshed by the zeros of $S_N(x)$ defined in (2). Substituting each x_k into (4), it can be expressed (in matrix form) as

$$\begin{bmatrix} u(x_1) \\ u(x_2) \\ \vdots \\ u(x_N) \end{bmatrix} = \begin{bmatrix} S_0(x_1) & S_1(x_1) & \cdots & S_{N-1}(x_1) \\ S_0(x_2) & S_1(x_2) & \cdots & S_{N-1}(x_2) \\ \vdots & \vdots & \ddots & \vdots \\ S_0(x_N) & S_1(x_N) & \cdots & S_{N-1}(x_N) \end{bmatrix} \begin{bmatrix} c_0 \\ c_1 \\ \vdots \\ c_{N-1} \end{bmatrix},$$

which is denoted by $\mathbf{u} = \mathbf{S}\mathbf{c}$. The unknown coefficient vector can be performed by $\mathbf{c} = \mathbf{S}^{-1}\mathbf{u}$. Let us consider the single-layer integration of $u(x)$ from 0 to x_k, which is denoted by $U^{(1)}(x_k)$; we obtain

$$U^{(1)}(x_k) = \int_0^{x_k} u(\xi)\,d\xi = \sum_{n=0}^{N-1} c_n \int_0^{x_k} S_n(\xi)\,d\xi = \sum_{n=0}^{N-1} c_n \bar{S}_n(x_k)$$

for $k \in \{1,2,3,\ldots,N\}$ or, in matrix form,

$$\begin{bmatrix} U^{(1)}(x_1) \\ U^{(1)}(x_2) \\ \vdots \\ U^{(1)}(x_N) \end{bmatrix} = \begin{bmatrix} \bar{S}_0(x_1) & \bar{S}_1(x_1) & \cdots & \bar{S}_{N-1}(x_1) \\ \bar{S}_0(x_2) & \bar{S}_1(x_2) & \cdots & \bar{S}_{N-1}(x_2) \\ \vdots & \vdots & \ddots & \vdots \\ \bar{S}_0(x_N) & \bar{S}_1(x_N) & \cdots & \bar{S}_{N-1}(x_N) \end{bmatrix} \begin{bmatrix} c_0 \\ c_1 \\ \vdots \\ c_{N-1} \end{bmatrix}.$$

We denote the above matrix by $\mathbf{U}^{(1)} = \bar{\mathbf{S}}\mathbf{c} = \bar{\mathbf{S}}\mathbf{S}^{-1}\mathbf{u} := \mathbf{A}\mathbf{u}$, where $\mathbf{A} = \bar{\mathbf{S}}\mathbf{S}^{-1} := [a_{ki}]_{N \times N}$ is called the first-order shifted Chebyshev integration matrix for the FIM-SCP; that is,

$$U^{(1)}(x_k) = \int_0^{x_k} u(\xi)\,d\xi = \sum_{i=1}^N a_{ki} u(x_i).$$

Next, consider the double-layer integration of $u(x)$ from 0 to x_k, which denoted by $U^{(2)}(x_k)$. We have

$$U^{(2)}(x_k) = \int_0^{x_k} \int_0^{\xi_2} u(\xi_1) d\xi_1 d\xi_2 = \sum_{i=1}^{N} a_{ki} \int_0^{x_i} u(\xi_1) d\xi_1 = \sum_{i=1}^{N} \sum_{j=1}^{N} a_{ki} a_{ij} u(x_j)$$

for $k \in \{1, 2, 3, ..., N\}$. It can be written, in matrix form, as $\mathbf{U}^{(2)} = \mathbf{A}^2 \mathbf{u}$. Similarly, we can calculate the n-layer integration of $u(x)$ from 0 to x_k, which is denoted by $U^{(n)}(x_k)$. Then, we have

$$U^{(n)}(x_k) = \int_0^{x_k} \cdots \int_0^{\xi_2} u(\xi_1) d\xi_1 \cdots d\xi_n = \sum_{i_n=1}^{N} \cdots \sum_{j=1}^{N} a_{k i_n} \cdots a_{i_1 j} u(x_j)$$

for $k \in \{1, 2, 3, ..., N\}$, which can be expressed, in matrix form, as $\mathbf{U}^{(n)} = \mathbf{A}^n \mathbf{u}$.

2.2. Tikhonov Regularization Method

In this section, we briefly present the idea of the Tikhonov regularization method [19], which is usually applied to stabilize ill-posed problems, such as our inverse problem. Normally, the considered inverse problem can be represented by the system of m linear equations with n unknowns, as

$$\mathbf{A}\mathbf{x} = \mathbf{b}^\epsilon, \tag{5}$$

where $\mathbf{b}^\epsilon$ is the vector in the right-hand side, which is perturbed by some noise ϵ, and $\mathbf{x}$ is the solution of the system (5) after perturbation. Tikhonov regularization replaces the inverse problem (5) by a minimization problem to obtain an efficiently approximate solution, which can be described as

$$\arg\min_{\mathbf{x} \in \mathbb{R}^n} \left\{ \|\mathbf{A}\mathbf{x} - \mathbf{b}^\epsilon\|^2 + \lambda \|\mathbf{x}\|^2 \right\}, \tag{6}$$

where $\lambda > 0$ is a regularization parameter balancing the weighting between the two terms of the function and $\|\cdot\|$ is the standard Euclidean norm. To reformulate the above minimization problem (6), we obtain

$$\arg\min_{\mathbf{x} \in \mathbb{R}^n} \left\{ \left\| \begin{bmatrix} \mathbf{A} \\ \sqrt{\lambda}\mathbf{I} \end{bmatrix} \mathbf{x} - \begin{bmatrix} \mathbf{b}^\epsilon \\ 0 \end{bmatrix} \right\|^2 \right\}.$$

Clearly, this is a linear least-square problem in $\mathbf{x}$. Then, the above problem turns out to be the normal equation of the form

$$\begin{bmatrix} \mathbf{A} \\ \sqrt{\lambda}\mathbf{I} \end{bmatrix}^\top \begin{bmatrix} \mathbf{A} \\ \sqrt{\lambda}\mathbf{I} \end{bmatrix} \mathbf{x} = \begin{bmatrix} \mathbf{A} \\ \sqrt{\lambda}\mathbf{I} \end{bmatrix}^\top \begin{bmatrix} \mathbf{b}^\epsilon \\ 0 \end{bmatrix}.$$

To simplify the above equation, the solution $\mathbf{x}$ under the regularization parameter λ (denoted by $\mathbf{x}_\lambda$) can be computed by

$$\mathbf{x}_\lambda = (\mathbf{A}^\top \mathbf{A} + \lambda \mathbf{I})^{-1} \mathbf{A}^\top \mathbf{b}^\epsilon. \tag{7}$$

We can see that the accuracy of $\mathbf{x}_\lambda$ in (7) depends on the regularization parameter λ, which plays an important role in the calculation: A large regularization parameter may over-smoothen the solution, while a small regularization parameter may lose the ability to stabilize the solution. Therefore, a suitable choice of the regularization parameter λ is very significant for finding a stable approximate solution. There are many approaches for choosing a value of the parameter λ, such as the discrepancy principle criterion, the generalized cross-validation, the L-curve method, and so on. Nevertheless, the regularization parameter λ in this paper is chosen according to Morozov's discrepancy principle combined with Newton's method, which has been proposed in Reference [20]. We provide the procedure for calculating the optimal regularization parameter λ below, which can be carried out by the following steps:

Step 1: Set $n = 0$ and give an initial regularization parameter $\lambda_0 > 0$.

Step 2: Compute $\mathbf{x}_{\lambda_n} = (\mathbf{A}^\top \mathbf{A} + \lambda_n \mathbf{I})^{-1} \mathbf{A}^\top \mathbf{b}^\epsilon$.

Step 3: Compute $\nabla \mathbf{x}_{\lambda_n} = -(\mathbf{A}^\top \mathbf{A} + \lambda_n \mathbf{I})^{-1} \mathbf{x}_{\lambda_n}$.

Step 4: Compute $G(\lambda_n) = \|\mathbf{A}\mathbf{x}_{\lambda_n} - \mathbf{b}^\epsilon\|^2 - \epsilon^2$.

Step 5: Compute $G'(\lambda_n) = 2\lambda_n \|\mathbf{A}\nabla\mathbf{x}_{\lambda_n}\|^2 + 2\lambda_n^2 \|\nabla \mathbf{x}_{\lambda_n}\|^2$.

Step 6: Compute $\lambda_{n+1} = \lambda_n - \frac{G(\lambda_n)}{G'(\lambda_n)}$.

Step 7: If $\|\lambda_{n+1} - \lambda_n\| < \delta$ for a tolerance δ, end. Else, set $n = n+1$ and return to Step 2.

Therefore, we receive the optimal regularization parameter λ, which is the terminal value λ_n obtained from the above procedure. When the regularization parameter λ is fixed as the mentioned optimal value, we can directly obtain the corresponding regularized solution by (7).

3. Numerical Algorithms for Direct and Inverse Problems of TVIDE

In this section, we apply the FIM-SCP described in Section 2.1 to devise the numerical algorithms for solving both the direct and inverse TVIDE problems (1), in order to obtain accurate approximate results. Let u be an approximate solution of v in (1). Then, we have the following linear TVIDE over the domain $\Omega = (0, L) \times (0, T)$:

$$u_t(x,t) + \mathcal{L}u(x,t) = \int_0^t \kappa_1(x,\eta) u(x,\eta) d\eta + \int_0^x \kappa_2(\xi,t) u(\xi,t) d\xi + F(x,t), \tag{8}$$

subject to the initial condition

$$u(x,0) = \phi(x), \quad x \in [0, L], \tag{9}$$

and the boundary conditions

$$u^{(r)}(b,t) = \psi_r(t), \quad t \in [0, T], \tag{10}$$

for $b \in \{0, L\}$ and $r \in \{0, 1, 2, ..., n-1\}$, where t and x represent time and space variables, respectively. Additionally, κ_1, κ_2, F, ϕ, and ψ_r are given continuous functions and $\mathcal{L}$ is the spatial linear differential operator of order n defined by $\mathcal{L} := \sum_{i=0}^{n} p_i(x,t) \frac{d^i}{dx^i}$, where $p_i(x,t)$ is given and sufficiently smooth.

3.1. Procedure for Solving the Direct TVIDE Problem

First, we linearize (8) by uniformly discretizing the temporal domain into M subintervals with time step τ. Then, we specify (8) at a time $t_m = m\tau$ for $m \in \mathbb{N}$ and use the first-order forward difference quotient to estimate the time derivative term u_t. Next, we replace each x by x_k for $k \in \{1, 2, 3, ..., N\}$ as generated by the zeros of the shifted Chebyshev polynomial $S_N(x)$ defined in (2). Thus, we have

$$\frac{u^{\langle m \rangle} - u^{\langle m-1 \rangle}}{\tau} + \mathcal{L}u^{\langle m \rangle} = \int_0^{t_m} \kappa_1(x_k, \eta) u(x_k, \eta) d\eta + \int_0^{x_k} \kappa_2(\xi, t_m) u(\xi, t_m) d\xi + F^{\langle m \rangle}, \tag{11}$$

where $u^{\langle m \rangle} = u^{\langle m \rangle}(x_k) = u(x_k, t_m)$ and $F^{\langle m \rangle} = F^{\langle m \rangle}(x_k) = F(x_k, t_m)$. Next, consider the first integral term with respect to time by letting it be $J_1^{\langle m \rangle}(x_k)$, we approximate $J_1^{\langle m \rangle}(x_k)$ by using the trapezoidal rule. Thus, we approximate $J_1^{\langle m \rangle}(x_k)$ as

$$J_1^{\langle m\rangle}(x_k) := \int_0^{t_m} \kappa_1(x_k,\eta)u(x_k,\eta)d\eta$$

$$= \sum_{i=0}^{m-1} \int_{t_i}^{t_{i+1}} \kappa_1(x_k,\eta)u(x_k,\eta)d\eta$$

$$\approx \sum_{i=0}^{m-1} \frac{\tau}{2}\left(\kappa_1^{\langle i\rangle}(x_k)u^{\langle i\rangle}(x_k) + \kappa_1^{\langle i+1\rangle}(x_k)u^{\langle i+1\rangle}(x_k)\right)$$

$$= \frac{\tau}{2}\kappa_1^{\langle 0\rangle}(x_k)u^{\langle 0\rangle}(x_k) + \tau\sum_{i=1}^{m-1} \kappa_1^{\langle i\rangle}(x_k)u^{\langle i\rangle}(x_k) + \frac{\tau}{2}\kappa_1^{\langle m\rangle}(x_k)u^{\langle m\rangle}(x_k)$$

for each $x_k \in \{x_1, x_2, x_3, ..., x_N\}$. The above equation can be written, in matrix form, as

$$\mathbf{J}_1^{\langle m\rangle} = \frac{\tau}{2}\mathbf{K}_1^{\langle 0\rangle}\mathbf{u}^{\langle 0\rangle} + \tau\sum_{i=1}^{m-1} \mathbf{K}_1^{\langle i\rangle}\mathbf{u}^{\langle i\rangle} + \frac{\tau}{2}\mathbf{K}_1^{\langle m\rangle}\mathbf{u}^{\langle m\rangle}, \tag{12}$$

where each parameter in (12) can be defined as follows:

$$\begin{aligned}
\mathbf{J}_1^{\langle m\rangle} &= \left[J_1^{\langle m\rangle}(x_1), J_1^{\langle m\rangle}(x_2), J_1^{\langle m\rangle}(x_3), ..., J_1^{\langle m\rangle}(x_N)\right]^\top, \\
\mathbf{u}^{\langle i\rangle} &= \left[u^{\langle i\rangle}(x_1), u^{\langle i\rangle}(x_2), u^{\langle i\rangle}(x_3), ..., u^{\langle i\rangle}(x_N)\right]^\top, \\
\mathbf{K}_1^{\langle i\rangle} &= \text{diag}\left(\kappa_1^{\langle i\rangle}(x_1), \kappa_1^{\langle i\rangle}(x_2), \kappa_1^{\langle i\rangle}(x_3), ..., \kappa_1^{\langle i\rangle}(x_N)\right).
\end{aligned}$$

Then, we consider the second integral term with respect to space by letting it be $J_2^{\langle m\rangle}(x_k)$ and using the idea of FIM-SCP (as described in Section 2.1) to approximate it. Then, we obtain

$$J_2^{\langle m\rangle}(x_k) := \int_0^{x_k} \kappa_2(\xi, t_m)u(\xi, t_m)d\xi = \int_0^{x_k} \kappa_2^{\langle m\rangle}(\xi)u^{\langle m\rangle}(\xi)d\xi \approx \sum_{i=1}^N a_{ki}\kappa_2^{\langle m\rangle}(x_i)u^{\langle m\rangle}(x_i)$$

for each $x_k \in \{x_1, x_2, x_3, ..., x_N\}$. The above equation can be written, in matrix form, as

$$\mathbf{J}_2^{\langle m\rangle} = \mathbf{A}\mathbf{K}_2^{\langle m\rangle}\mathbf{u}^{\langle m\rangle}, \tag{13}$$

where $\mathbf{A} = \bar{\mathbf{S}}\mathbf{S}^{-1}$ is the shifted Chebyshev integration matrix defined in Section 2.1,

$$\begin{aligned}
\mathbf{J}_2^{\langle m\rangle} &= \left[J_2^{\langle m\rangle}(x_1), J_2^{\langle m\rangle}(x_2), J_2^{\langle m\rangle}(x_3), ..., J_2^{\langle m\rangle}(x_N)\right]^\top, \\
\mathbf{u}^{\langle m\rangle} &= \left[u^{\langle m\rangle}(x_1), u^{\langle m\rangle}(x_2), u^{\langle m\rangle}(x_3), ..., u^{\langle m\rangle}(x_N)\right]^\top, \\
\mathbf{K}_2^{\langle m\rangle} &= \text{diag}\left(\kappa_2^{\langle m\rangle}(x_1), \kappa_2^{\langle m\rangle}(x_2), \kappa_2^{\langle m\rangle}(x_3), ..., \kappa_2^{\langle m\rangle}(x_N)\right).
\end{aligned}$$

Then, we apply the FIM-SCP (described in Section 2.1) to eliminate all spatial derivatives from (11) by taking the n-layer integral on both sides of (11), to obtain the following equation at the shifted Chebyshev node x_k, as defined in (2), as

$$\int_0^{x_k}\cdots\int_0^{\xi_2}\left(\frac{u^{\langle m\rangle} - u^{\langle m-1\rangle}}{\tau} + \mathcal{L}u^{\langle m\rangle}\right)d\xi_1...d\xi_n = \int_0^{x_k}\cdots\int_0^{\xi_2}\left(J_1^{\langle m\rangle} + J_2^{\langle m\rangle} + F^{\langle m\rangle}\right)d\xi_1...d\xi_n. \tag{14}$$

To simplify the n-layer integration of the spatial derivative terms of $\mathcal{L}u^{\langle m \rangle}$, by letting it be $Q^{\langle m \rangle}(x_k)$ and using the technique of integration by parts, we have

$$Q^{\langle m \rangle}(x_k) := \int_0^{x_k} \cdots \int_0^{\xi_2} \mathcal{L}u^{\langle m \rangle}(\xi_1) d\xi_1 \ldots d\xi_n$$

$$= \int_0^{x_k} \cdots \int_0^{\xi_2} \sum_{i=0}^n p_i(\xi_1, t_m) \frac{d^i}{dx^i} u^{\langle m \rangle}(\xi_1) d\xi_1 \ldots d\xi_n$$

$$= \sum_{i=0}^n (-1)^i \binom{n}{i} \int_0^{x_k} \cdots \int_0^{\eta_2} p_n^{(i)}(\eta_1, t_m) u^{\langle m \rangle}(\eta_1) d\eta_1 \ldots d\eta_i$$

$$+ \int_0^{x_k} \left(\sum_{i=0}^{n-1} (-1)^i \binom{n-1}{i} \int_0^{\xi_n} \cdots \int_0^{\eta_2} p_{n-1}^{(i)}(\eta_1, t_m) u^{\langle m \rangle}(\eta_1) d\eta_1 \ldots d\eta_i \right) d\xi_n$$

$$+ \int_0^{x_k} \int_0^{\xi_n} \left(\sum_{i=0}^{n-2} (-1)^i \binom{n-2}{i} \int_0^{\xi_{n-1}} \cdots \int_0^{\eta_2} p_{n-2}^{(i)}(\eta_1, t_m) u^{\langle m \rangle}(\eta_1) d\eta_1 \ldots d\eta_i \right) d\xi_{n-1} d\xi_n$$

$$\vdots$$

$$+ \int_0^{x_k} \cdots \int_0^{\xi_2} p_0(\xi_1, t_m) u^{\langle m \rangle}(\xi_1) d\xi_1 \ldots d\xi_n + d_1 \frac{x_k^{n-1}}{(n-1)!} + d_2 \frac{x_k^{n-2}}{(n-2)!} + d_3 \frac{x_k^{n-3}}{(n-3)!} + \ldots + d_n,$$

where $d_1, d_2, d_3, \ldots, d_n$ are the arbitrary constants which emerge from the process of integration by parts. Then, we substitute each $x_k \in \{x_1, x_2, x_3, \ldots, x_N\}$ into the above equation and utilize the idea of FIM-SCP. Thus, we can express it, in matrix form, by

$$\mathbf{Q}^{\langle m \rangle} = \sum_{i=0}^n (-1)^i \binom{n}{i} \mathbf{A}^i \mathbf{P}_n^{(i)} \mathbf{u}^{\langle m \rangle} + \sum_{i=0}^{n-1} (-1)^i \binom{n-1}{i} \mathbf{A}^{i+1} \mathbf{P}_{n-1}^{(i)} \mathbf{u}^{\langle m \rangle}$$

$$+ \sum_{i=0}^{n-2} (-1)^i \binom{n-2}{i} \mathbf{A}^{i+2} \mathbf{P}_{n-2}^{(i)} \mathbf{u}^{\langle m \rangle} + \cdots + \mathbf{A}^n \mathbf{P}_0^{(0)} \mathbf{u}^{\langle m \rangle} + \mathbf{X}_n \mathbf{d} \quad (15)$$

$$= \sum_{j=0}^n \sum_{i=0}^{n-j} (-1)^i \binom{n-j}{i} \mathbf{A}^{i+j} \mathbf{P}_{n-j}^{(i)} \mathbf{u}^{\langle m \rangle} + \mathbf{X}_n \mathbf{d},$$

where $\mathbf{A} = \bar{\mathbf{S}} \mathbf{S}^{-1}$ is the shifted Chebyshev integration matrix, $\mathbf{d} = [d_1, d_2, d_3, \ldots, d_N]^\top$,

$$\mathbf{Q}^{\langle m \rangle} = \left[Q^{\langle m \rangle}(x_1), Q^{\langle m \rangle}(x_2), Q^{\langle m \rangle}(x_3), \ldots, Q^{\langle m \rangle}(x_N) \right]^\top,$$

$$\mathbf{X}_n = [\mathbf{x}_{n-1}, \mathbf{x}_{n-2}, \mathbf{x}_{n-3}, \ldots, \mathbf{x}_0] \text{ for each } \mathbf{x}_i = \frac{1}{i!} [x_1^i, x_2^i, x_3^i, \ldots, x_N^i]^\top,$$

$$\mathbf{P}_{n-j}^{(i)} = \text{diag}\left(p_{n-j}^{(i)}(x_1, t_m), p_{n-j}^{(i)}(x_2, t_m), p_{n-j}^{(i)}(x_3, t_m), \ldots, p_{n-j}^{(i)}(x_N, t_m) \right).$$

Finally, we vary all points $x_k \in \{x_1, x_2, x_3, \ldots, x_N\}$ in (14) and rearrange them into matrix form by using the FIM-SCP with the derived matrix equations (12), (13), and (15); thus, we obtain

$$\frac{\mathbf{A}^n \mathbf{u}^{\langle m \rangle} - \mathbf{A}^n \mathbf{u}^{\langle m-1 \rangle}}{\tau} + \mathbf{Q}^{\langle m \rangle} = \mathbf{A}^n \mathbf{J}_1^{\langle m \rangle} + \mathbf{A}^n \mathbf{J}_2^{\langle m \rangle} + \mathbf{A}^n \mathbf{F}^{\langle m \rangle}$$

or, factorizing the unknown solution $\mathbf{u}^{\langle m \rangle}$ explicitly, as

$$\left(\mathbf{A}^n + \tau \sum_{j=0}^n \sum_{i=0}^{n-j} (-1)^i \binom{n-j}{i} \mathbf{A}^{i+j} \mathbf{P}_{n-j}^{(i)} - \frac{\tau^2}{2} \mathbf{A}^n \mathbf{K}_1^{\langle m \rangle} - \tau \mathbf{A}^{n+1} \mathbf{K}_2^{\langle m \rangle} \right) \mathbf{u}^{\langle m \rangle}$$
$$+ \mathbf{X}_n \mathbf{d} = \frac{\tau^2}{2} \mathbf{A}^n \mathbf{K}_1^{\langle 0 \rangle} \mathbf{u}^{\langle 0 \rangle} + \tau^2 \sum_{i=1}^{m-1} \mathbf{A}^n \mathbf{K}_1^{\langle i \rangle} \mathbf{u}^{\langle i \rangle} + \mathbf{A}^n \mathbf{u}^{\langle m-1 \rangle} + \tau \mathbf{A}^n \mathbf{F}^{\langle m \rangle}. \quad (16)$$

Next, consider the given boundary conditions (10) at the endpoints $b \in \{0, L\}$. We can convert them into matrix form by using the linear combination of shifted Chebyshev polynomial (4) in term of the rth-order derivative of u at the iteration time t_m and using (3). Then, we have

$$\frac{d^r}{dx^r} u^{\langle m \rangle}(x) \Big|_{x=b} = \sum_{n=0}^{N-1} c_n^{\langle m \rangle} \frac{d^r}{dx^r} S_n(x) \Big|_{x=b} = \psi_r(t_m)$$

for all $r \in \{0, 1, 2, ..., n-1\}$. We can express the above equation, in matrix form, as

$$\begin{bmatrix} S_0(b) & S_1(b) & \cdots & S_{N-1}(b) \\ S_0'(b) & S_1'(b) & \cdots & S_{N-1}'(b) \\ \vdots & \vdots & \ddots & \vdots \\ S_0^{(n-1)}(b) & S_1^{(n-1)}(b) & \cdots & S_{N-1}^{(n-1)}(b) \end{bmatrix} \begin{bmatrix} c_0^{\langle m \rangle} \\ c_1^{\langle m \rangle} \\ \vdots \\ c_{N-1}^{\langle m \rangle} \end{bmatrix} = \begin{bmatrix} \psi_0(t_m) \\ \psi_1(t_m) \\ \vdots \\ \psi_{n-1}(t_m) \end{bmatrix}, \quad (17)$$

which can be denoted by $\mathbf{B}\mathbf{c}^{\langle m \rangle} = \mathbf{\Psi}^{\langle m \rangle}$ or $\mathbf{B}\mathbf{S}^{-1}\mathbf{u}^{\langle m \rangle} = \mathbf{\Psi}^{\langle m \rangle}$. Finally, we can construct the system of mth iterative linear equations from (16) and (17), which has $N + n$ unknowns containing $\mathbf{u}^{\langle m \rangle}$ and $\mathbf{d}$, as follows:

$$\begin{bmatrix} \mathbf{H}^{\langle m \rangle} & \mathbf{X}_n \\ \mathbf{B}\mathbf{S}^{-1} & \mathbf{0} \end{bmatrix} \begin{bmatrix} \mathbf{u}^{\langle m \rangle} \\ \mathbf{d} \end{bmatrix} = \begin{bmatrix} \mathbf{E}_1^{\langle m \rangle} \\ \mathbf{\Psi}^{\langle m \rangle} \end{bmatrix}, \quad (18)$$

where $\mathbf{H}^{\langle m \rangle}$ is the coefficient matrix of $\mathbf{u}^{\langle m \rangle}$ in (16) and $\mathbf{E}_1^{\langle m \rangle}$ is the right-hand side column vector of (16). Consequently, the solution $\mathbf{u}^{\langle m \rangle}$ can be approximated by solving the system (18) starting from the given initial condition (9); that is, $\mathbf{u}^{\langle 0 \rangle} = [\phi(x_1), \phi(x_2), \phi(x_3), ..., \phi(x_N)]^\top$. Note that, when we would like to find a numerical solution $u(x, t)$ at any point $x \in [0, L]$ for the terminal time T, we can calculate it by the following formula:

$$u(x, T) = \sum_{n=0}^{N-1} c_n^{\langle m \rangle} S_n(x) = \mathbf{s}(x)\mathbf{c}^{\langle m \rangle} = \mathbf{s}(x)\mathbf{S}^{-1}\mathbf{u}^{\langle m \rangle},$$

where $\mathbf{s}(x) = [S_0(x), S_1(x), S_2(x), ..., S_{N-1}(x)]$ and $\mathbf{u}^{\langle m \rangle}$ is the final m^{th} iterative solution of (18).

3.2. Procedure for Solving Inverse Problem of TVIDE

For the inverse problem in this paper, we specifically define the forcing term $F(x,t) := \beta(t)f(x,t)$, where $\beta(t)$ is a missing source function to be retrieved and $f(x,t)$ is the given function. Thus, our considered time-dependent inverse TVIDE problem (1) becomes

$$u_t(x,t) + \mathcal{L}u(x,t) = \int_0^t \kappa_1(x,\eta)u(x,\eta)d\eta + \int_0^x \kappa_2(\xi,t)u(\xi,t)d\xi + \beta(t)f(x,t), \quad (19)$$

where u is an approximate solution of v and the other parameters are defined as in (8). The initial and boundary conditions of (19) are (9) and (10), which satisfy the compatibility conditions. Now, we remove all spatial derivatives from (19) and use the shifted Chebyshev integration matrix (as explained in Section 2.1). Then, we obtain the following matrix equation, based on the same process as in (16), as

$$\left(\mathbf{A}^n + \tau \sum_{j=0}^{n} \sum_{i=0}^{n-j}(-1)^i \binom{n-j}{i} \mathbf{A}^{i+j} \mathbf{P}_{n-j}^{(i)} - \frac{\tau^2}{2}\mathbf{A}^n \mathbf{K}_1^{\langle m \rangle} - \tau \mathbf{A}^{n+1}\mathbf{K}_2^{\langle m \rangle} \right) \mathbf{u}^{\langle m \rangle} \\ + \mathbf{X}_n \mathbf{d} - \tau \mathbf{A}^n \mathbf{f}^{\langle m \rangle} \beta^{\langle m \rangle} = \frac{\tau^2}{2} \mathbf{A}^n \mathbf{K}_1^{\langle 0 \rangle} \mathbf{u}^{\langle 0 \rangle} + \tau^2 \sum_{i=1}^{m-1} \mathbf{A}^n \mathbf{K}_1^{\langle i \rangle} \mathbf{u}^{\langle i \rangle} + \mathbf{A}^n \mathbf{u}^{\langle m-1 \rangle}, \quad (20)$$

where $\beta^{\langle m \rangle} = \beta(t_m)$, $\mathbf{f}^{\langle m \rangle} = [f(x_1, t_m), f(x_2, t_m), f(x_3, t_m), ..., f(x_N, t_m)]^\top$ and the other parameters in (20) are as defined in Section 3.1. However, the occurrence of missing data is caused by the given conditions being insufficient to ensure a unique solution to our inverse problem. Hence, an additional condition or observed data needs to be involved. Thus, we use an additional condition, regarding the aggregated solution of the system, in the following form:

$$\int_0^L u(\xi,t)\,d\xi = g(t), \quad t \in [0,T], \tag{21}$$

where $g(t)$ is the measured data at time t, which probably contains measurement errors. In order to illustrate the realistic phenomena of this problem, we assume that the measurement data of the aggregated solution $g(t)$ involves some noise ϵ, which is denoted by $g^\epsilon(t)$ (where $\|g^\epsilon(t) - g(t)\| \leq \epsilon$) and define the noisy value ϵ by a random variable generated by the Gaussian normal distribution with mean $\mu = 0$ and standard deviation $\sigma = p|g(t)|$, where p is the percentage of the noise to be input. Then, the additional condition (21) becomes

$$\int_0^L u(\xi,t)\,d\xi = g^\epsilon(t), \quad t \in [0,T]. \tag{22}$$

Using the concept of FIM-SCP, the additional condition (22) at time t_m can be written, in vector form, as

$$\int_0^L u^{\langle m\rangle}(\xi)\,d\xi = \sum_{n=0}^{N-1} c_n^{\langle m\rangle} \int_0^L S_n(\xi)\,d\xi = \sum_{n=0}^{N-1} c_n^{\langle m\rangle} \bar{S}_n(L)\,d\xi := \mathbf{zc}^{\langle m\rangle} = \mathbf{zS}^{-1}\mathbf{u}^{\langle m\rangle} = g^\epsilon(t_m), \tag{23}$$

where $\mathbf{z} = [\bar{S}_0(L), \bar{S}_1(L), \bar{S}_2(L), ..., \bar{S}_{N-1}(L)]$ and each $\bar{S}_n(L)$ is as defined in Lemma 1(iii). Finally, we can establish the following system of mth iterative linear equations for the inverse TVIDE problem (19) by utilizing (20) and (23), which has $N + n + 1$ unknown variables including $\mathbf{u}^{\langle m\rangle}$, $\mathbf{d}$, and $\beta^{\langle m\rangle}$, as

$$\begin{bmatrix} \mathbf{H}^{\langle m\rangle} & \mathbf{X}_n & -\tau\mathbf{A}^n\mathbf{f}^{\langle m\rangle} \\ \mathbf{BS}^{-1} & 0 & 0 \\ \mathbf{zS}^{-1} & 0 & 0 \end{bmatrix} \begin{bmatrix} \mathbf{u}^{\langle m\rangle} \\ \mathbf{d} \\ \beta^{\langle m\rangle} \end{bmatrix} = \begin{bmatrix} \mathbf{E}_2^{\langle m\rangle} \\ \mathbf{\Psi}^{\langle m\rangle} \\ g^\epsilon(t_m) \end{bmatrix}, \tag{24}$$

where $\mathbf{H}^{\langle m\rangle}$ is the coefficient matrix of $\mathbf{u}^{\langle m\rangle}$ defined in (20) and $\mathbf{E}_2^{\langle m\rangle}$ is the right-hand side column vector of (20). Before seeking an approximate solution $\mathbf{u}^{\langle m\rangle}$ and source term $\beta^{\langle m\rangle}$, as we have mentioned, we must address that our inverse problem is ill-posed. When a noisy value is input into the system, it may cause a significant error. Hence, we need to stabilize the solution of (24) by employing the Tikhonov regularization method. We denote the linear system (24) by the simplified matrix equation as

$$\mathbf{Ry} = \mathbf{b}^\epsilon. \tag{25}$$

Applying the Tikhonov regularization method (6) in order to filter out the noise in the corresponding perturbed data, we can stabilize the numerical solution (25) by using (7). Thus, we have

$$\mathbf{y}_\lambda = (\mathbf{R}^\top\mathbf{R} + \lambda\mathbf{I})^{-1}\mathbf{R}^\top\mathbf{b}^\epsilon. \tag{26}$$

Finally, we can receive the optimal regularization parameter λ by using Morozov's discrepancy principle combined with Newton's method, as described in Section 2.2. Thus, we can directly obtain the corresponding regularized solution by (26).

3.3. Algorithms for Solving the Direct and Inverse TVIDE Problems

For computational convenience, we summarize the aforementioned procedures for finding approximate solutions to the direct (8) and inverse (19) TVIDE problems in Sections 3.1 and 3.2, respectively, as the numerical Algorithms 1 and 2, which are in the form of pseudocode.

Algorithm 1 Numerical algorithm for solving the direct TVIDE problem via FIM-SCP

Input: x, τ, L, T, N, $\phi(x)$, $\psi_r(t)$, $p_i(x,t)$, $\kappa_1(x,t)$, $\kappa_2(x,t)$, and $F(x,t)$.

Output: An approximate solution $u(x,T)$.

1: Set $x_k = \frac{L}{2}\left(\cos\left(\frac{2k-1}{2N}\pi\right)+1\right)$ for $k \in \{1,2,3,...,N\}$ in descending order.
2: Compute $\mathbf{A}$, $\mathbf{B}$, $\mathbf{S}$, $\bar{\mathbf{S}}$, $\mathbf{S}^{-1}$, $\mathbf{X}_n$, and $\mathbf{u}^{\langle 0 \rangle}$.
3: Set $m = 1$ and $t_1 = \tau$.
4: **while** $t_m \leq T$ **do**
5:      Compute $\mathbf{K}_1^{\langle m \rangle}$, $\mathbf{K}_2^{\langle m \rangle}$, $\mathbf{F}^{\langle m \rangle}$, $\mathbf{H}^{\langle m \rangle}$, $\mathbf{\Psi}^{\langle m \rangle}$, and $\mathbf{E}_1^{\langle m \rangle}$.
6:      Find $\mathbf{u}^{\langle m \rangle}$ by solving the linear system (18).
7:      Update $m = m + 1$.
8:      Compute $t_m = m\tau$.
9: **end while**
10: **return** Find $u(x,T) = \mathbf{s}(x)\mathbf{S}^{-1}\mathbf{u}^{\langle m \rangle}$.

Algorithm 2 Numerical algorithm for solving the inverse TVIDE problem via FIM-SCP

Input: x, p, τ, δ, L, T, N, λ_0, $\phi(x)$, $g(t)$, $\psi_r(t)$, $p_i(x,t)$, $\kappa_1(x,t)$, $\kappa_2(x,t)$, and $f(x,t)$.

Output: An approximate solution $u(x,T)$ and the source terms $\beta(t_m)$ at all discretized times.

1: Set $x_k = \frac{L}{2}\left(\cos\left(\frac{2k-1}{2N}\pi\right)+1\right)$ for $k \in \{1,2,3,...,N\}$ in descending order.
2: Compute $\mathbf{A}$, $\mathbf{B}$, $\mathbf{S}$, $\bar{\mathbf{S}}$, $\mathbf{S}^{-1}$, $\mathbf{X}_n$, $\mathbf{A}_N$, and $\mathbf{u}^{\langle 0 \rangle}$.
3: Set $m = 1$ and $t_1 = \tau$.
4: **while** $t_m \leq T$ **do**
5:      Set the measurement data $g^\epsilon(t_m) = g(t_m) + \epsilon$, where $\epsilon \sim \mathcal{N}(0, p^2|g(t_m)|^2)$.
6:      Compute $\mathbf{K}_1^{\langle m \rangle}$, $\mathbf{K}_2^{\langle m \rangle}$, $\mathbf{f}^{\langle m \rangle}$, $\mathbf{H}^{\langle m \rangle}$, $\mathbf{\Psi}^{\langle m \rangle}$, $\mathbf{E}_2^{\langle m \rangle}$, $\mathbf{R}$, and $\mathbf{b}^\epsilon$.
7:      Set $n = 0$.
8:      **do**
9:          Compute $\mathbf{y}_{\lambda_n} = (\mathbf{R}^\top \mathbf{R} + \lambda_n \mathbf{I})^{-1}\mathbf{R}^\top \mathbf{b}^\epsilon$.
10:        Compute $\nabla \mathbf{y}_{\lambda_n} = -(\mathbf{R}^\top \mathbf{R} + \lambda_n \mathbf{I})^{-1}\mathbf{y}_{\lambda_n}$.
11:        Compute $G(\lambda_n) = \|\mathbf{R}\mathbf{y}_{\lambda_n} - \mathbf{b}^\epsilon\|^2 - \epsilon^2$.
12:        Compute $G'(\lambda_n) = 2\lambda_n\|\mathbf{R}\nabla\mathbf{y}_{\lambda_n}\|^2 + 2\lambda_n^2\|\nabla\mathbf{y}_{\lambda_n}\|^2$.
13:        Compute $\lambda_{n+1} = \lambda_n - \frac{G(\lambda_n)}{G'(\lambda_n)}$.
14:        Update $n = n + 1$.
15:      **while** $\|\lambda_n - \lambda_{n-1}\| \geq \delta$
16:      Set the optimal regularization parameter $\lambda = \lambda_n$.
17:      Find $\mathbf{u}^{\langle m \rangle}$ and $\beta^{\langle m \rangle}$ by explicitly solving $\mathbf{y}_\lambda$ using the matrix equation (7).
18:      Update $m = m + 1$.
19:      Compute $t_m = m\tau$.
20: **end while**
21: **return** Find $u(x,T) = \mathbf{s}(x)\mathbf{S}^{-1}\mathbf{u}^{\langle m \rangle}$.

4. Numerical Experiments

In this section, we implement our devised numerical algorithms for solving the direct and inverse TVIDE problems through several examples, in order to demonstrate the efficiency and accuracy of the solutions obtained by proposed methods. Examples 1 and 2 are used to examine Algorithm 1 for the direct TVIDE problems (8). Examples 3 and 4 are inverse TVIDE problems (19), as solved by Algorithm 2. Additionally, time convergence rates and CPU times(s) for each example are presented to indicate the computational cost and time. The time convergence rate is defined by Rate = $\lim_{t_m \to T} \frac{\|\mathbf{u}^*(t_{m+1}) - \mathbf{u}(t_{m+1})\|_\infty}{\|\mathbf{u}^*(t_m) - \mathbf{u}(t_m)\|_\infty}$, where T is the terminal time, t_m is a partitioned time contained in $[0, T]$, $\mathbf{u}^*(t_m)$ is the exact solution at time t_m, $\mathbf{u}(t_m)$ is the numerical solution at time t_m, and $\|\cdot\|_\infty$ is the l^∞ norm. Graphical solutions of each example are also depicted. Our numerical algorithms were implemented using the MatLab R2016a software, run on a Intel(R) Core(TM) i7-6700 CPU @ 3.40 GHz computer system.

Example 1. *Consider the following direct TVIDE problem, which consists of a second-order derivative with constant coefficient for $x \in (0,1)$ and $t \in (0,T]$:*

$$u_t + u_{xx} + u = \int_0^t 2e^{-x} u(x,\eta) d\eta + \int_0^x (\xi + t) u(\xi, t) d\xi + F(x,t), \tag{27}$$

where

$$F(x,t) = -t - e^x(t^2 - 3t + tx - 1),$$

subject to the homogeneous initial condition $u(x,0) = 0$ for $x \in [0,1]$ and the Dirichlet boundary conditions $u(0,t) = t$ and $u(1,t) = te$ for $t \in [0,T]$. The analytical solution of this problem is $u^*(x,t) = te^x$.

In the numerical testing based on Algorithm 1, we first took the double-layer integral of both sides of (27) and transformed it into matrix form (16). Then, we obtained the approximate solutions $u(x,T)$ for this problem (27) by applying the numerical Algorithm 1. The accuracy of our obtained approximate results was measured by the mean absolute error, which compared it to the analytical solution at different values of $x \in \{0.1, 0.3, 0.5, 0.7, 0.9\}$ and the terminal time $T = 1$, as shown in Table 1. From Table 1, we observe that, when the partitioning number of the temporal domain M was fixed and nodal numbers N were increasingly varied, then the accuracy was significantly improved. Similarly, for a fixed nodal number N but various time partitioning numbers M, the accuracy results were also significantly improved. Moreover, the convergence rates with respect to the time in Algorithm 1 were estimated for various numbers of the time partition ($M \in \{5, 10, 15, 20, 25\}$) for the spatial points $N = 10$, as shown in Table 2. We can notice, from Table 2, that these time convergence rates for the ℓ^∞ norm indeed approached linear convergence for $T \in \{5, 10, 15\}$. The computational cost, in terms of CPU times (s), is also displayed in Table 2. Finally, a graph of our approximate solutions $u(x,t)$ for different times t and the surface plot of the solution under the parameters $N = 20$, $M = 20$, and $T = 1$ are depicted in Figure 1.

Table 1. Mean absolute errors between exact and numerical solutions of $u(x,1)$ for Example 1.

x	$M = 20$			$N = 12$		
	$N = 8$	$N = 10$	$N = 12$	$M = 11$	$M = 13$	$M = 15$
0.1	1.6855×10^{-5}	1.1723×10^{-8}	1.0208×10^{-10}	2.3823×10^{-7}	5.0060×10^{-9}	3.3268×10^{-11}
0.3	4.3851×10^{-5}	3.0501×10^{-8}	2.6554×10^{-10}	6.1976×10^{-7}	1.3024×10^{-8}	8.6539×10^{-11}
0.5	5.3554×10^{-5}	3.7247×10^{-8}	3.2429×10^{-10}	7.5684×10^{-7}	1.5904×10^{-8}	1.0567×10^{-10}
0.7	4.2555×10^{-5}	2.9599×10^{-8}	2.5767×10^{-10}	6.0140×10^{-7}	1.2638×10^{-8}	8.3964×10^{-11}
0.9	1.5864×10^{-5}	1.1034×10^{-8}	9.6049×10^{-11}	2.2419×10^{-7}	4.7112×10^{-9}	3.1289×10^{-11}

Table 2. Time convergence rates and CPU times (s) for Example 1 by Algorithm 1 with $N = 10$.

M	$T = 5$			$T = 10$			$T = 15$		
	$\|\|u^* - u\|\|_\infty$	Rate	Time(s)	$\|\|u^* - u\|\|_\infty$	Rate	Time(s)	$\|\|u^* - u\|\|_\infty$	Rate	Time(s)
5	4.298×10^{-12}	1.4584	0.0465	8.565×10^{-12}	1.5076	0.0456	1.262×10^{-11}	1.5112	0.0469
10	4.318×10^{-12}	1.2499	0.0487	8.576×10^{-12}	1.2863	0.0469	1.276×10^{-11}	1.3040	0.0485
15	4.309×10^{-12}	1.1723	0.0495	8.547×10^{-12}	1.2008	0.0481	1.275×10^{-11}	1.2135	0.0501
20	4.311×10^{-12}	1.1327	0.0506	8.533×10^{-12}	1.1555	0.0516	1.277×10^{-11}	1.1657	0.0535
25	1.135×10^{-12}	1.1353	0.0521	8.540×10^{-12}	1.1272	0.0538	1.277×10^{-11}	1.1365	0.0553

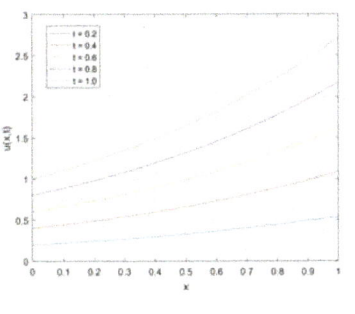
(a) $u(x, t)$ at different times t

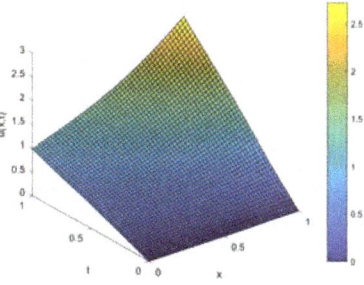
(b) Surface plot of $u(x, t)$

Figure 1. The graphical results of Example 1 for $N = 20$, $M = 20$, and $T = 1$.

Example 2. *Consider the following direct TVIDE problem, which consists of a third-order derivative with variable coefficient for $x \in (0, 1)$ and $t \in (0, T]$:*

$$u_t + tu_{xxx} + \cos(x)u_{xx} = \int_0^t \frac{2}{x-1} u(x, \eta) d\eta + \int_0^x \frac{6t}{\xi - 1} u(\xi, t) d\xi + F(x, t), \quad (28)$$

where

$$F(x, t) = x - x^2 + xt^2(3x + 1) - 2t\cos(x),$$

subject to the initial condition $u(x, 0) = 0$ for $x \in [0, 1]$ and the boundary conditions $u(0, t) = 0$, $u(1, t) = 0$, and $u'(0, t) = t$ for $t \in [0, T]$. The analytical solution of this problem is $u^*(x, t) = (x - x^2)t$.

We test the efficiency and accuracy of the proposed Algorithm 1 via the problem (28). First, we took a triple-layer integral on both sides of (28) and utilized the shifted Chebyshev integration matrix to transform it into matrix form (16). Next, we implemented Algorithm 1 to obtain numerical solutions $u(x, T)$ for this problem (28). Table 3 shows the precision of our obtained approximate results at different values of $x \in \{0.1, 0.3, 0.5, 0.7, 0.9\}$ and at the terminal time $T = 1$, through the mean absolute error. We can see that the accuracy was significantly improved according to an increase in the number of both the partitioning space and time domains. However, we observe that, in the case of fixed N, when M was increased, the mean absolute errors provide accurate results with a lower computational number M. Furthermore, the time convergence rates concerning the ℓ^∞ norm and CPU times (s) are demonstrated in Table 4, under various values of M ($M \in \{5, 10, 15, 20, 25\}$) and final times T ($T \in \{5, 10, 15\}$). The graphical solutions for $u(x, t)$ in both one and two dimensions are shown in Figure 2.

Table 3. Mean absolute errors between exact and numerical solutions of $u(x,1)$ for Example 2.

x	M = 10			N = 10		
	N = 8	N = 10	N = 12	M = 5	M = 10	M = 15
0.1	3.5098×10^{-10}	5.0498×10^{-13}	1.6695×10^{-14}	5.1849×10^{-13}	5.0498×10^{-13}	4.9435×10^{-13}
0.3	1.0060×10^{-9}	1.1285×10^{-12}	4.1411×10^{-14}	1.1544×10^{-12}	1.1285×10^{-12}	1.0850×10^{-12}
0.5	1.0780×10^{-9}	1.4543×10^{-12}	5.1958×10^{-14}	1.4672×10^{-12}	1.4543×10^{-12}	1.3845×10^{-12}
0.7	1.0237×10^{-9}	1.1625×10^{-12}	4.5908×10^{-14}	1.1572×10^{-12}	1.1625×10^{-12}	1.0923×10^{-12}
0.9	3.1567×10^{-10}	5.2400×10^{-13}	2.0983×10^{-14}	5.1567×10^{-13}	5.2400×10^{-13}	4.9050×10^{-13}

Table 4. Time convergence rates and CPU times (s) for Example 2 by Algorithm 1 with $N = 10$.

M	T = 5			T = 10			T = 15		
	$\|u^* - u\|_\infty$	Rate	Time(s)	$\|u^* - u\|_\infty$	Rate	Time(s)	$\|u^* - u\|_\infty$	Rate	Time(s)
5	1.426×10^{-12}	1.0241	0.0524	1.620×10^{-12}	1.0489	0.0531	3.549×10^{-12}	1.3720	0.0535
10	1.533×10^{-12}	1.0334	0.0577	1.635×10^{-12}	1.0278	0.0576	1.874×10^{-12}	1.1035	0.0576
15	1.426×10^{-12}	1.0208	0.0597	1.664×10^{-12}	1.0243	0.0585	1.806×10^{-12}	1.0294	0.0598
20	1.476×10^{-12}	1.0223	0.0609	1.609×10^{-12}	1.0210	0.0610	1.537×10^{-12}	1.0203	0.0620
25	1.496×10^{-12}	1.0165	0.0619	1.488×10^{-12}	1.0182	0.0638	1.276×10^{-12}	1.0079	0.0641

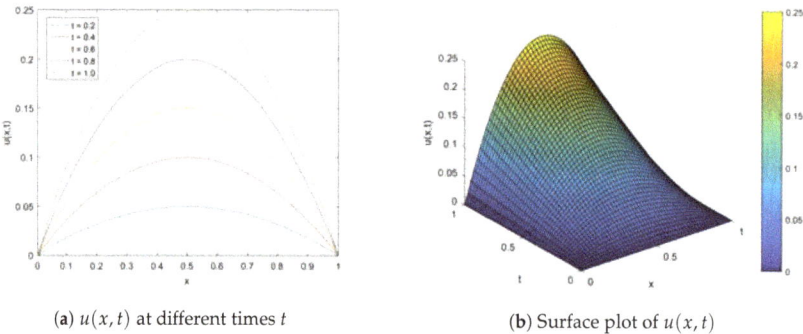

(a) $u(x,t)$ at different times t (b) Surface plot of $u(x,t)$

Figure 2. The graphical results of Example 2 for $N = 20$, $M = 20$, and $T = 1$.

Example 3. *Consider the following inverse TVIDE problem, which consists of a second-order derivative with constant coefficient and a continuous forcing function $f(x,t)$ for $x \in (0,1)$ and $t \in (0,T]$:*

$$u_t - u_{xx} + 2u = \int_0^t 2\ln(x)u(x,\eta)d\eta + \int_0^x e^{-\xi}u(\xi,t)d\xi + \beta(t)f(x,t), \tag{29}$$

where

$$f(x,t) = e^{2t}\left[1 + t - x + e^x + te^{-x} - (2e^x + t)t\ln x\right],$$

subject to the initial condition $u(x,0) = e^x$ for $x \in [0,1]$ and the boundary conditions $u(0,t) = t + 1$ and $u(1,t) = t + e$ for $t \in [0,T]$. The additional condition, in terms of the aggregated solution of the system, is $g(t) = t + e - 1$. The analytical solutions of this problem are $u^*(x,t) = t + e^x$ and $\beta^*(t) = e^{-2t}$.

Implementing the numerical Algorithm 2 by taking the double-layer integral of both sides of (29) and transforming it into matrix form (24), we obtained the approximate solutions $u(x,1)$ and $\beta(t)$ for this problem (29). As the additional condition was measurement data, there may be an error in the measurement. Therefore, we perturbed the additional condition $g(t)$ with a percentage p of the noise ($p \in \{0\%, 1\%, 3\%, 5\%\}$). In Table 5, we show the accuracy of the solutions $u(x,1)$ and $\beta(t)$, in terms of the mean absolute error, respectively, denoted by $\mathcal{E}_u = \frac{1}{N}\sum_{i=1}^N |u_i^* - u_i|$ and $\mathcal{E}_\beta = \frac{1}{M}\sum_{j=1}^M |\beta_j^* - \beta_j|$, and the values of the optimal regularization parameters λ at time $t = 1$ with various $M = N \in$

{5, 10, 15, 20}. From Table 5, we can observe that the optimal regularization parameters λ were close to zero and the mean absolute errors for both $\mathcal{E}_u$ and $\mathcal{E}_\beta$ significantly increased with an increasing percentage p of the perturbation. Furthermore, we used the regularization parameter $\lambda = 0$ to explore the rates of convergence with respect to the ℓ^∞ norm and CPU times (s) for $M = N \in \{5, 10, 15, 20\}$ with the final times $T \in \{1, 2, 3\}$ as shown in Table 6. The graphical solutions of the perturbed functions $u(x, 1)$ and $\beta(t)$ for $p \in \{1\%, 3\%, 5\%\}$ are depicted in Figure 3.

Table 5. Mean absolute errors of $u(x,1)$ and $\beta(t)$ for optimal regularization parameter λ of Example 3.

$M=N$	$p=0\%$			$p=1\%$		
	λ	$\mathcal{E}_u$	$\mathcal{E}_\beta$	λ	$\mathcal{E}_u$	$\mathcal{E}_\beta$
5	6.22×10^{-14}	1.6609×10^{-5}	7.9997×10^{-7}	3.11×10^{-12}	1.1372×10^{-4}	4.1098×10^{-4}
10	2.38×10^{-18}	3.9844×10^{-13}	1.0459×10^{-12}	2.20×10^{-13}	3.4011×10^{-4}	8.8853×10^{-4}
15	1.02×10^{-17}	9.9950×10^{-14}	1.6384×10^{-13}	9.17×10^{-12}	7.8857×10^{-4}	7.0288×10^{-4}
20	2.14×10^{-18}	2.9774×10^{-13}	1.7125×10^{-13}	4.33×10^{-14}	1.9024×10^{-4}	1.3201×10^{-3}
$M=N$	$p=3\%$			$p=5\%$		
	λ	$\mathcal{E}_u$	$\mathcal{E}_\beta$	λ	$\mathcal{E}_u$	$\mathcal{E}_\beta$
5	8.80×10^{-11}	1.2533×10^{-3}	8.3870×10^{-3}	8.61×10^{-12}	2.2805×10^{-3}	1.0964×10^{-2}
10	6.64×10^{-11}	3.7067×10^{-3}	9.5414×10^{-3}	1.11×10^{-11}	5.0047×10^{-3}	2.7003×10^{-2}
15	1.09×10^{-12}	8.4361×10^{-3}	7.0094×10^{-3}	8.79×10^{-12}	3.2925×10^{-2}	3.2201×10^{-2}
20	8.61×10^{-12}	3.3382×10^{-3}	1.3582×10^{-2}	1.40×10^{-13}	1.0214×10^{-2}	3.8774×10^{-2}

Table 6. Time convergence rates and CPU times (s) for Example 3 by Algorithm 2 with $N = 10$.

M	$T=1$			$T=2$			$T=3$		
	$\|u^* - u\|_\infty$	Rate	Time(s)	$\|u^* - u\|_\infty$	Rate	Time(s)	$\|u^* - u\|_\infty$	Rate	Time(s)
5	8.495×10^{-13}	1.0046	0.0667	9.130×10^{-13}	1.0203	0.0655	3.602×10^{-12}	1.6644	0.0662
10	8.131×10^{-13}	0.9970	0.0679	8.659×10^{-13}	1.0042	0.0667	2.456×10^{-12}	1.2409	0.0673
15	7.851×10^{-13}	0.9965	0.0684	8.362×10^{-13}	1.0001	0.0675	1.731×10^{-12}	1.1355	0.0693
20	8.344×10^{-13}	1.0003	0.0716	7.829×10^{-13}	0.9967	0.0722	2.928×10^{-12}	1.0956	0.0720
25	8.362×10^{-13}	1.0003	0.0776	8.686×10^{-13}	1.0022	0.0766	2.134×10^{-12}	1.0699	0.0751

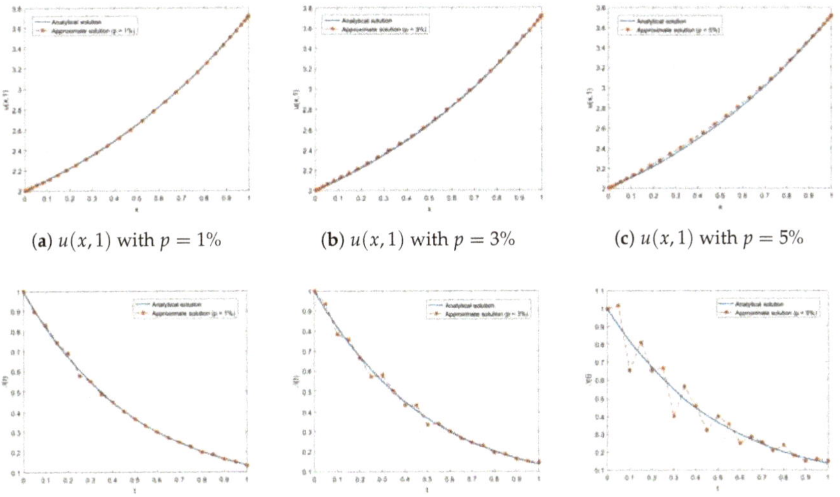

(a) $u(x,1)$ with $p=1\%$

(b) $u(x,1)$ with $p=3\%$

(c) $u(x,1)$ with $p=5\%$

(d) $\beta(t)$ with $p=1\%$

(e) $\beta(t)$ with $p=3\%$

(f) $\beta(t)$ with $p=5\%$

Figure 3. The graphical results of $u(x,1)$ and $\beta(t)$ for Example 3 with $N = 30$ and $M = 20$.

Example 4. Consider the following inverse TVIDE problem, which consists of a second-order derivative with variable coefficient and the piecewise forcing function $f(x,t)$ for $x \in (0,1)$ and $t \in (0,T]$:

$$u_t + u_{xx} + u_x - \cos(xt)u = \int_0^t 2\sin(x)u(x,\eta)d\eta - \int_0^x 3t\cos(\xi)u(\xi,t)d\xi + \beta(t)f(x,t), \quad (30)$$

where

$$f(x,t) = \begin{cases} \frac{1}{2}\left[2t\cos(2x) + t\sin(2x) + (t\cos(xt)+1)\sin^2 x\right], & 0 < t \leq \frac{T}{3}, \\ \frac{1}{3}\left[2t\cos(2x) + t\sin(2x) + (t\cos(xt)+1)\sin^2 x\right], & \frac{T}{3} < t \leq \frac{2T}{3}, \\ \frac{1}{4}\left[2t\cos(2x) + t\sin(2x) + (t\cos(xt)+1)\sin^2 x\right], & \frac{2T}{3} < t \leq T, \end{cases}$$

subject to the initial condition $u(x,0) = 0$ for $x \in [0,1]$ and the Dirichlet boundary conditions $u(0,t) = 0$ and $u(1,t) = t\sin^2(1)$ for $t \in [0,T]$. The additional condition, in terms of the aggregated solution of the system, is $g(t) = \frac{t}{4}(2 - \sin(2)) + e^t$. The analytical solutions of this problem are $u^*(x,t) = t\sin^2(x)$ and

$$\beta^*(t) = \begin{cases} 2, & 0 < t \leq \frac{T}{3}, \\ 3, & \frac{T}{3} < t \leq \frac{2T}{3}, \\ 4, & \frac{2T}{3} < t \leq T. \end{cases}$$

Based on the numerical Algorithm 2, we took the double-layer integral to both sides of (30) and transformed it into matrix form (24). We obtained the approximate solutions $u(x,1)$ and $\beta(t)$ for (29) by implementing Algorithm 2. Table 7 shows the accuracy of the solutions $u(x,1)$ and $\beta(t)$ obtained by our numerical algorithm, in terms of the mean absolute errors $\mathcal{E}_u$ and $\mathcal{E}_\beta$, as well as the values of the optimal regularization parameter λ at time $t = 1$ with the noisy percentage $p \in \{0\%, 1\%, 5\%, 10\%\}$ under various $M = N \in \{6, 9, 12, 15\}$. Although this problem had the piecewise forcing term $f(x,t)$, our Algorithm 2 perfectly performed in providing accurate results, as shown in Table 7. The time convergence rates concerning the ℓ^∞ norm and CPU times (s) are shown in Table 8, under various numbers of $M \in \{6, 9, 12, 15, 18\}$ with the final times $T \in \{1, 2, 3\}$. The graphical perturbed solutions $u(x,1)$ and $\beta(t)$ for $p \in \{1\%, 5\%, 10\%\}$ are shown in Figure 4.

Table 7. Mean absolute errors of $u(x,1)$ and $\beta(t)$ for optimal regularization parameter λ of Example 4.

$M = N$	$p = 0\%$			$p = 1\%$		
	λ	$\mathcal{E}_u$	$\mathcal{E}_\beta$	λ	$\mathcal{E}_u$	$\mathcal{E}_\beta$
6	1.25×10^{-14}	6.4987×10^{-6}	18123×10^{-4}	2.60×10^{-12}	7.8604×10^{-6}	1.8959×10^{-4}
9	7.41×10^{-17}	2.8057×10^{-9}	7.1782×10^{-9}	3.45×10^{-10}	1.2932×10^{-7}	4.0319×10^{-5}
12	2.65×10^{-20}	1.4031×10^{-13}	4.5672×10^{-12}	6.02×10^{-11}	2.6701×10^{-7}	5.1531×10^{-5}
15	6.11×10^{-21}	5.4903×10^{-14}	2.7330×10^{-13}	8.41×10^{-12}	2.9871×10^{-6}	6.8381×10^{-5}
$M = N$	$p = 5\%$			$p = 10\%$		
	λ	$\mathcal{E}_u$	$\mathcal{E}_\beta$	λ	$\mathcal{E}_u$	$\mathcal{E}_\beta$
6	4.19×10^{-12}	8.8690×10^{-5}	5.1802×10^{-4}	5.51×10^{-11}	6.5680×10^{-4}	3.7335×10^{-3}
9	6.20×10^{-13}	1.8419×10^{-5}	7.0830×10^{-4}	7.96×10^{-11}	4.4035×10^{-4}	2.3292×10^{-3}
12	4.11×10^{-12}	7.0910×10^{-5}	1.6889×10^{-3}	8.65×10^{-12}	6.4815×10^{-4}	6.3981×10^{-3}
15	1.01×10^{-13}	2.7507×10^{-4}	1.7821×10^{-3}	5.64×10^{-12}	5.9709×10^{-4}	6.1579×10^{-3}

Table 8. Time convergence rates and CPU times (s) for Example 4 by Algorithm 2 with $N = 12$.

M	$T=1$			$T=2$			$T=3$		
	$\|u^* - u\|_\infty$	Rate	Time(s)	$\|u^* - u\|_\infty$	Rate	Time(s)	$\|u^* - u\|_\infty$	Rate	Time(s)
6	2.681×10^{-13}	1.4574	0.0704	5.338×10^{-13}	1.4564	0.0726	8.024×10^{-13}	1.4563	0.0728
9	2.677×10^{-13}	1.3415	0.0727	5.353×10^{-13}	1.3406	0.0737	8.006×10^{-13}	1.3395	0.0746
12	2.681×10^{-13}	1.2758	0.0735	5.338×10^{-13}	1.2744	0.0753	8.011×10^{-13}	1.2743	0.0767
15	2.666×10^{-13}	1.2312	0.0749	5.360×10^{-13}	1.2332	0.0783	8.015×10^{-13}	1.2337	0.0807
18	2.682×10^{-13}	1.2028	0.0798	5.351×10^{-13}	1.2031	0.0799	7.989×10^{-13}	1.2022	0.0828

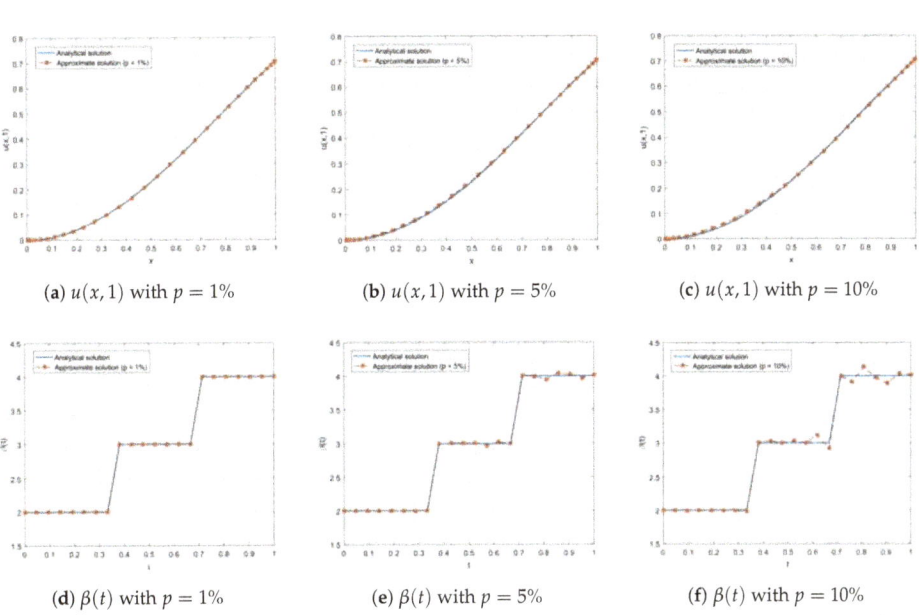

(a) $u(x,1)$ with $p = 1\%$ (b) $u(x,1)$ with $p = 5\%$ (c) $u(x,1)$ with $p = 10\%$

(d) $\beta(t)$ with $p = 1\%$ (e) $\beta(t)$ with $p = 5\%$ (f) $\beta(t)$ with $p = 10\%$

Figure 4. The graphical results $u(x,T)$ and $\beta(t)$ for Example 4 with $N = 30$ and $M = 21$.

5. Conclusions and Discussion

In this paper, we utilized FIM-SCP combined with the forward difference quotient to create efficient and accurate numerical algorithms for solving the considered direct and inverse TVIDE problems. According to the numerical examples in Section 4, we have demonstrated the performance of our proposed Algorithm 1 for seeking the approximate solutions of direct TVIDE problems in Examples 1 and 2. We can see that, for Example 1—which involved a second-order derivative with constant coefficients—Algorithm 1 provided an accurate result. Furthermore, for a problem involving a higher-order derivative with variable coefficients, it still provided high accuracy, in terms of solutions, as demonstrated in Example 2. Moreover, we handled inverse TVIDE problems using Algorithm 2, the effectiveness of which was illustrated in Examples 3 and 4. We used the Tikhonov regularization method to deal with the instability of the inverse problem; it can be seen that, in the examples, the regularization parameter λ was close to zero. Algorithm 2 could handle both continuous and piecewise-defined forcing terms with high accuracy, as demonstrated in Examples 3 and 4. Furthermore, when we perturbed the problems by adding noisy values, our Algorithm 2 still overcame the noise and provided approximate results that approached the analytical solutions. We further notice that our presented methods provide high accuracy, even when using only a small number of nodal points. Evidently, when we decrease the time step, they will furnish more accurate results. The rates of convergence with respect to time (based on the ℓ^∞ norm) of our methods were observed to be linear.

Finally, we also depicted the computational times for each example. However, we realize that there exist no theoretical error analysis results for the proposed numerical algorithms. Thus, our future research will study the error analysis, in order to find theories for order of accuracy and rate of convergence for our method. Another interesting direction for our future work is to extend our techniques to solve other types of IDEs and non-linear IDEs.

Author Contributions: Conceptualization, R.B., A.D., and P.G.; methodology, R.B. and A.D.; software, A.D. and P.G.; validation, R.B., A.D., and P.G.; formal analysis, R.B.; investigation, A.D. and P.G.; writing—original draft preparation, A.D. and P.G.; writing—review and editing, R.B.; visualization, A.D. and P.G.; supervision, R.B.; project administration, R.B.; funding acquisition, R.B. All authors have read and agreed to the published version of the manuscript.

Acknowledgments: The authors would like to thank the reviewers for their thoughtful comments and efforts towards improving our manuscript.

Conflicts of Interest: The authors declare no conflict of interest.

Abbreviations

The following abbreviations are used in this manuscript:

FDM	finite difference method
FIM	finite integration method
FIM-SCP	finite integration method with shifted Chebyshev polynomial
IDE	integro-differential equation
PDE	partial differential equation
TVIDE	time-dependent Volterra integro-differential equation

References

1. Zill, D.G.; Wright, W.S.; Cullen, M.R. *Differential Equations with Boundary-Value Problem*, 8th ed.; Brooks/Cole, Cengang Learning: Boston, MA, USA, 2013.
2. Yanik, E.G.; Fairweather, G. Finite element methods for parabolic and hyperbolic partial integro–differential equations. *Nonlinear Anal.* **1988**, *12*, 785–809. [CrossRef] [CrossRef]
3. Engle, H. *On Some Parabolic Integro–Differential Equations: Existence and Asymptotics of Solution*; Lecture Notes in Mathematics, Springer: Berlin, Germany, 1983.
4. Tang, T. A finite difference scheme for partial integro–differential equations with a weakly singular kernel. *Appl. Numer. Math.* **1993**, *11*, 309–319. [CrossRef] [CrossRef]
5. Aguilar, M.; Brunner, H. Collocation methods for second–order Volterra integro–differential equations. *Appl. Numer. Math.* **1988**, *4*, 455–470. [CrossRef] [CrossRef]
6. Brunner, H. Implicit Runge–Kutta–Nyström methods for general second–order Volterra integro–differential equations. *Comput. Math. Appl.* **1987**, *14*, 549–559. [CrossRef] [CrossRef]
7. Jiang, Y.J. On spectral methods for Volterra-type integro–differential equations. *J. Comput. Appl. Math.* **2009**, *230*, 333–340. [CrossRef] [CrossRef]
8. Burton, T.A. *Volterra Integral and Differential Equations*; Academic Press: New York, NY, USA, 1983.
9. Rahman, M. *Integral Equations and Their Applications*; WIT Press: Southampton, UK, 2007.
10. Hu, Q. Stieltjes derivatives and beta–polynomial spline collocation for Volterra integro–differential equations with singularities. *SIAM J. Numer.* **1996**, *33*, 208–220. [CrossRef] [CrossRef]
11. Brunner, H. Superconvergence in collocation and implicit Runge–Kutta methods for Volterra–type integral equations of the second kind. *Internet Schriftenreihe Numer. Math.* **1980**, *53*, 54–72. [CrossRef]
12. El-Sayed, S.M.; Kaya, D.; Zarea, S. The decomposition method applied to solve high–order linear Volterra–Fredholm integro–differential equations. *Internet J. Nonlinear Sci. Numer. Simulat.* **2004**, *5*, 105–112. [CrossRef] [CrossRef]
13. Kabanikhin, S.I. Definitions and examples of inverse and ill–posed problems. *J. Inverse Ill-Pose Probl.* **2008**, *16*, 317–357. [CrossRef] [CrossRef]
14. Wen, P.H.; Hon, Y.C.; Li, M.; Korakianitis, T. Finite integration method for partial differential equations. *Appl. Math. Model.* **2013**, *37*, 10092–10106. [CrossRef] [CrossRef]

15. Li, M.; Chen, C.S.; Hon, Y.C.; Wen, P.H. Finite integration method for solving multi–dimensional partial differential equations. *Appl. Math. Model.* **2015**, *39*, 4979–4994. [CrossRef] [CrossRef]
16. Li, M.; Tian, Z.L.; Hon, Y.C.; Chen, C.S.; Wen, P.H. Improved finite integration method for partial differential equations. *Eng. Anal. Bound. Elem.* **2016**, *64*, 230–236. [CrossRef] [CrossRef]
17. Boonklurb, R.; Duangpan, A.; Treeyaprasert, T. Modified finite integration method using Chebyshev polynomial for solving linear differential equations. *J. Numer. Ind. Appl. Math.* **2018**, *12*, 1–19. [CrossRef]
18. Rivlin, T.J. *Chebyshev Polynomials, From Approximation Theory to Algebra and Number Theory*, 2nd ed.; John Wiley and Sons: New York, NY, USA, 1990.
19. Tikhonov, A.N.; Goncharsky, A.V.; Stepanov, V.V.; Yagola, A.G. *Numerical Methods for the Solution of Ill–Posed Problems*; Springer: Dordrecht, The Netherlands, 1995. [CrossRef]
20. Sun, Y. Indirect boundary integral equation method for the Cauchy problem of the Laplace equation. *J. Sci. Comput.* **2017**, *71*, 469–498. [CrossRef] [CrossRef]

© 2020 by the authors. Licensee MDPI, Basel, Switzerland. This article is an open access article distributed under the terms and conditions of the Creative Commons Attribution (CC BY) license (http://creativecommons.org/licenses/by/4.0/).

Article

Numerical Solution of the Navier–Stokes Equations Using Multigrid Methods with HSS-Based and STS-Based Smoothers

Galina Muratova [1,*,†,‡], **Tatiana Martynova** [1,†,‡], **Evgeniya Andreeva** [1,†], **Vadim Bavin** [1,†] and **Zeng-Qi Wang** [2]

1. Mechanics and Computer Science, Vorovich Institute of Mathematics, Southern Federal University, Rostov-on-Don 344000, Russia; martynova@sfedu.ru (T.M.); andreeva@sfedu.ru (E.A.); tefp9999@mail.ru (V.B.)
2. School of Mathematical Sciences, Shanghai Jiao Tong University, Shanghai 200000, China; wangzengqi@sjtu.edu.cn
* Correspondence: muratova@sfedu.ru; Tel.: +7-863-219-9736
† Current address: 200/1 Stachki Ave., Bld. 2, Rostov-on-Don 344090, Russia.
‡ These authors contributed equally to this work.

Received: 29 November 2019; Accepted: 21 January 2020; Published: 4 February 2020

Abstract: Multigrid methods (MGMs) are used for discretized systems of partial differential equations (PDEs) which arise from finite difference approximation of the incompressible Navier–Stokes equations. After discretization and linearization of the equations, systems of linear algebraic equations (SLAEs) with a strongly non-Hermitian matrix appear. Hermitian/skew-Hermitian splitting (HSS) and skew-Hermitian triangular splitting (STS) methods are considered as smoothers in the MGM for solving the SLAE. Numerical results for an algebraic multigrid (AMG) method with HSS-based smoothers are presented.

Keywords: multigrid methods; Hermitian/skew-Hermitian splitting method; skew-Hermitian triangular splitting method; strongly non-Hermitian matrix

1. Introduction

Mathematical modeling of hydrodynamics is the base for research of various natural phenomena, technological processes, and environmental problems. The main equations describing this problem are the Navier–Stokes equations. Development and research of effective numerical algorithms for solving these equations and their practical realization is an actual task. The use of the MGM for the numerical solution of the Navier–Stokes equations describing the motion of an incompressible viscous fluid is discussed. Currently, various discretization methods for the corresponding differential model are known. However, with any choice of the discretizing method, the problem of constructing effective methods for solving large systems of algebraic equations—to which the discrete model is reduced—arises. This problem is especially relevant in the nonstationary case, when multiple solutions of the systems of algebraic equations are required at each discrete time step.

To discretize the system of two-dimensional Navier–Stokes equations on regular grids, we use the finite difference method. The equations are considered in the natural variables "velocity-pressure":

$$\frac{\partial \mathbf{V}}{\partial t} + (\mathbf{V} \cdot \nabla)\mathbf{V} = -\nabla P + \nu \Delta \mathbf{V}, \quad div \mathbf{V} = 0, \tag{1}$$

where P/ρ is replaced by P (i.e., ρ is normalized at 1), P is the static pressure, $\mathbf{V}$ is the velocity vector, and ν is the kinematic viscosity coefficient. At the initial moment of time and at the boundary of the domain, the initial and boundary conditions are set, respectively.

Flow simulating is accompanied by a number of mathematical difficulties. One of the problems in solving this system is the nonlinearity associated with convective terms in the equations, which can lead to the appearance of oscillations of the solution in regions with large gradients. The main efforts of the researchers were directed at overcoming the difficulties associated with the nonlinearity of the Navier–Stokes system of equations.

One of the most time-consuming stages of the computational procedure is finding the solution of the system of linear algebraic equations (SLAE). Modern application packages usually use the linearization of the original equations, and Krylov subspace methods are used to solve the resulting SLAEs. Despite the fact that these methods have proven themselves well, they have some problems in cases of significant nonsymmetry of the SLAEs—associated, for example, with variable coefficients in differential equations or using complex numerical boundary conditions. For time discretization of the unsteady problem, we use an implicit difference scheme. Here, we do not specifically consider the stages of discretization and linearization of the Navier–Stokes equations, but focus on solving SLAEs. Given that the SLAEs resulting from the use of the implicit time schemes have a large dimension and a sparse nonsymmetric matrix, we propose using the MGM to solve them.

Thus, we consider the iterative solution of the large sparse SLAE

$$Av = b, \quad v, b \in \mathbb{C}^n, \tag{2}$$

where $A \in \mathbb{C}^{n \times n}$ is a non-Hermitian and positive definite matrix.

Naturally, the matrix A can be split as

$$A = A_0 + A_1, \tag{3}$$

where

$$A_0 = \frac{1}{2}(A + A^*), \quad A_1 = \frac{1}{2}(A - A^*) \tag{4}$$

and A^* denotes the conjugate transpose of the matrix A. Positive definiteness of the matrix A means that for all $x \in \mathbb{C}^n \setminus \{0\}$, $x^* A_0 x > 0$. Here, x^* denotes the conjugate transpose of the complex vector x. Let in some matrix norm $|||\cdot|||$, $|||A_0||| << |||A_1|||$, then the matrix A is called a strongly non-Hermitian one. This situation occurs in many real applications, such as the discretization of the Navier–Stokes equations.

The Hermitian and skew-Hermitian splitting (HSS) iteration methods, based on HS splitting (3) and (4), for solving large sparse non-Hermitian positive definite SLAE were firstly proposed in [1]. The HSS iteration method has been widely developed in [2–5] and others.

Then, we can split the skew-Hermitian part A_1 of the matrix $A \in \mathbb{C}^{n \times n}$ into

$$A_1 = K_L + K_U, \tag{5}$$

where K_L and K_U are the strictly lower and the strictly upper triangular parts of A_1, respectively. Obviously, that $K_L = -K_U^*$.

Based on the splitting (3)–(5) in [6–8] classes of skew-Hermitian triangular splitting (STS), iteration methods for solving SLAE (2) have been proposed. The triangular operator of the STS uses only the skew-Hermitian part of the coefficient matrix A. These methods have been further developed in [9–12].

The use of the multigrid method (MGM) with the STS-based smoothers for solving convection–diffusion problems has been studied in [13]. The convergence of the MGM with the STS-based smoothers has also been proved in this research. The local Fourier analysis of the MGM with the triangular skew-symmetric smoothers has been performed in [14]. The results of numerical experiments for convection–diffusion problems with large Peclet numbers by the geometric MGM have been presented in both researches.

In [15], it was shown that the MGMs with the HSS-based smoothers converge uniformly for second-order nonselfadjoint elliptic boundary value problems. This happens if the mesh size of the coarsest grid is sufficiently small, but independent of the number of the multigrid levels.

2. Multigrid Methods

The MGMs are proving themselves as very successful tools for solving the SLAE associated with discretization of partial differential equations (PDEs).

The main idea of the MGM has been proposed by R.P. Fedorenko in [16]. Then, A. Brandt [17], W. Hackbusch [18], and other researchers showed the efficiency of the multigrid approach and extended Fedorenko's idea.

The multigrid technique is based on two principles: error smoothing and coarse grid correction. The smoothing property is fundamental for the MGM. It is connected with fast damping high-frequency Fourier components of an initial error in decomposition on the basis from eigenvectors.

There exist two approaches in the MGM: geometric multigrid and algebraic multigrid methods.

Geometric multigrid methods were critical to the early development of the MGM and still play an important role today. Nevertheless, there are classes of problems for which geometric techniques are too difficult to apply or cannot be used at all. These classes can be solved by the algebraic multigrid (AMG) methods, as introduced in [19,20].

The MGM is not a fixed algorithm. Rather, there is a multigrid technique that defines its scope. The efficiency of the MGM depends on the adjustment of its components to the considered problem [21]. The key to this is the correct choice of its components and effective interaction between smoothing and coarse-grid correction [22]. We need to use special iteration methods as smoothers for the MGM and nonstandard course-grid correction to a good approximation of the smooth error components.

The smoothing method is the central component of the multigrid algorithm; it is the most dependent part of the MGM on the type of the problem being solved. The role of smoothing methods is that they should not so much reduce the total error as smooth it (namely, suppress the high-frequency harmonics of the error) so that the error can be well approximated on a coarse grid.

Standard smoothing methods are linear iteration methods, for example, the Gauss–Seidel method. An alternative is the following methods:

- Richardson's Iterative method;
- Gauss–Jacobi method;
- Symmetric Gauss–Seidel method;
- Gauss–Seidel Alternate Direction method;
- Gauss–Seidel method with black and white ordering;
- Four-color Gauss–Seidel method;
- Iteration zebra method;
- Incomplete factorization method;
- Specially adapted SOR.

The MGMs can be used as solvers as well as preconditioners. The MGMs have been widely used for complicated nonsymmetric and nonlinear systems, like the Lame equations of elasticity or the Navier–Stokes problems.

3. Smoothers Based on the HSS and the STS Iteration Methods

A particular problem when using the MGM is the choice of smoothers. There are a number of iteration methods that can be used as smoothers, but not all of them are effective for solving strongly non-Hermitian SLAEs. The behavior of the HSS and the STS iteration methods is similar to the behavior of the Gauss–Seidel method, which quickly damps the high-frequency harmonics of the error, slowing down in the future. We give the formulas of these iteration methods.

The HSS iteration method [1]: Given an initial guess $v^{(0)}$, for $k = 0,1,2,...$ until $\{v^{(k)}\}$ convergence, compute

$$\begin{cases} (\alpha I + A_0)v^{(k+\frac{1}{2})} = (\alpha I - A_1)v^{(k)} + b, \\ (\alpha I + A_1)v^{(k+1)} = (\alpha I - A_0)v^{(k+\frac{1}{2})} + b, \end{cases}$$

where α is a given positive constant and I is an identity matrix.

Bai, Golub, and Ng [1] proved that the HSS iteration method converges unconditionally to the exact solution of the SLAE (2). Moreover, the upper bound of the contraction factor depends on the spectrum of A_0 but is independent of the spectrum of A_1.

We can rewrite the HSS iteration method in the following form:

$$v^{(k+1)} = G(\alpha)v^{(k)} + B(\alpha)^{-1}b,$$

where

$$G(\alpha) = B(\alpha)^{-1}(B(\alpha) - A)$$

and

$$B(\alpha) = \frac{1}{2\alpha}(\alpha I + A_0)(\alpha I + A_1).$$

The STS iteration method [6,8]: Given an initial guess $v^{(0)}$ and two positive parameters ω and τ. For $k = 0, 1, 2, \ldots$ until $\{v^{(k)}\}$ convergence, compute

$$v^{(k+1)} = G(\omega, \tau)v^{(k)} + \tau B(\omega)^{-1}b,$$

where

$$G(\omega, \tau) = B(\omega)^{-1}(B(\omega) - \tau A),$$

ω and τ are two acceleration parameters, and $B(\omega)$ is defined by

$$B(\omega) = B_c + \omega((1+j)K_L + (1-j)K_U), \quad j = \pm 1$$

with $B_c \in \mathbb{C}^{n \times n}$ a prescribed Hermitian matrix.

For the STS method a convergence analysis, optimal choice of parameters and an accelerating procedure have presented in [8]. As it was mentioned above, smoothers in the MGMs should have a smoothing effect on the error of approximation. It was shown in [14] that the skew-Hermitian triangular iteration methods have such properties. Therefore, these methods can be used as smoothers in the MGMs.

4. Numerical Experiments

A wide class of CFD (Computational Fluid Dynamics) problems is associated with solving the equations of motion of a viscous incompressible fluid with a predominance of convective transfer. As a model, we consider the problem of internal single-phase chemically homogeneous flows, which are described by the unsteady Navier–Stokes equations in the domain Ω with a solid boundary Γ. At the initial stages of the development of CFD, preference was given to explicit methods that were used to solve stationary and nonstationary Navier–Stokes equations. Recently, increased attention has been paid to implicit methods. This is primarily due to the insufficient computational efficiency of explicit methods in solving the equations of motion of a viscous fluid using small difference grids. From the point of view of computational linear algebra, the matrices obtained at each time step when integrating unsteady equations using implicit schemes (after linearization) are nonselfadjoint and require special iterative methods for their effective solution. Therefore, in this research, we suggest using the AMG with special smoothers to solve such SLAEs.

So, we consider the model unsteady Navier–Stokes problem

$$\frac{\partial \mathbf{V}}{\partial t} + (\mathbf{V} \cdot \nabla)\mathbf{V} = -\nabla P + \nu \Delta \mathbf{V}, \tag{6}$$

$$div\mathbf{V} = 0, \tag{7}$$

or

$$\frac{\partial u}{\partial t} + u\frac{\partial u}{\partial x} + v\frac{\partial u}{\partial y} + \frac{\partial P}{\partial x} - \frac{1}{Re}\left(\frac{\partial^2 u}{\partial x^2} + \frac{\partial^2 u}{\partial y^2}\right) = 0, \tag{8}$$

$$\frac{\partial v}{\partial t} + u\frac{\partial v}{\partial x} + v\frac{\partial v}{\partial y} + \frac{\partial P}{\partial y} - \frac{1}{Re}\left(\frac{\partial^2 v}{\partial x^2} + \frac{\partial^2 v}{\partial y^2}\right) = 0, \tag{9}$$

$$\frac{\partial u}{\partial x} + \frac{\partial v}{\partial y} = 0, \tag{10}$$

$$u(x,y,t) = 0, \quad v(x,y,t) = 0 \quad on \quad \Gamma,$$
$$u(x,y,0) = 0, \quad v(x,y,0) = 0,$$
$$P(x,y,0) = \xi x - \tfrac{1}{2}\xi, \quad \xi = const,$$

where ν is the kinematic viscosity coefficient; $Re = UL/\nu$ is the Reynolds number, where U is a characteristic velocity of the flow and L is a characteristic length scale; $\mathbf{V} = (u(x,y,t), v(x,y,t))$ is the velocity vector; P is the static pressure; the initial pressure distribution is given by a linear function. The initial conditions are taken to be zero. At the boundary, no-slip conditions are accepted. It means that at a solid boundary, the fluid will have zero velocity relative to the boundary. There are no mass forces in the formulation; motion is determined only by the boundary and initial conditions for the velocity field as well as the initial pressure distribution. For convenience, only square domain $\Omega = (0,1) \times (0,1)$ will be considered. We assume that the fluid motion occurs in the time interval $[0, T]$. Therefore, the equations are considered in the domain $\Omega \times (0, T)$ with the boundary $\Gamma \times [0, T]$. The Navier–Stokes equations with the introduced boundary conditions have a solution determined up to an arbitrary constant for pressure, therefore, an agreement was adopted on the next normalization $\int_\Omega P(x,y,t)dxdy - 0, \forall t$.

The most common approach to solving the Navier–Stokes equations in natural variables essentially uses the replacement of the difference continuity equation by the difference Poisson equation for pressure. Following this approach, first the difference equations are constructed that approximate the mass and momentum conservation equations and then, by algebraic transformations, the Poisson equation for determining the pressure is derived. This equation is used in the calculations instead of the continuity equation.

First, the equations of motion and continuity (6) and (7) are rewritten in schematic form [23]:

$$\frac{\partial \mathbf{V}}{\partial t} + \nabla P = \mathbf{R}, \tag{11}$$

where $\mathbf{R}$ contains all convective and diffusive forces,

$$\mathbf{R} = -(\mathbf{V} \cdot \nabla)\mathbf{V} + \frac{1}{Re}\Delta \mathbf{V}, \tag{12}$$

$$div\mathbf{V} = 0. \tag{13}$$

We fix the time step δt and introduce a discrete time grid $t^n = n\delta t$, $n \geq 0$ and denote the approximation to $f(t^n, x, y)$ as $f^{(n)}$. Then, the fully implicit scheme will have the form

$$\frac{1}{\delta t}(\mathbf{V}^{(n+1)} - \mathbf{V}^{(n)}) + \nabla P^{(n+1)} = \mathbf{R}^{(n+1)}, \tag{14}$$

$$\mathbf{R}^{(n+1)} = -(\mathbf{V}^{(n+1)} \cdot \nabla)\mathbf{V}^{(n+1)} + \frac{1}{Re}\Delta\mathbf{V}^{(n+1)}, \qquad (15)$$

$$div\mathbf{V}^{(n+1)} = 0, \qquad (16)$$

$$\mathbf{V}^{(n+1)}|_\Gamma = 0. \qquad (17)$$

The Poisson equation for pressure is obtained by taking the divergence from both sides of the Equation (14), taking into account the continuity Equation (16):

$$\Delta P^{(n+1)} - div\mathbf{R}^{(n+1)} = div\frac{\mathbf{V}^{(n)}}{\delta t}. \qquad (18)$$

But following [23], at this moment, the Poisson Equation (18) does not need to be created. Instead, we need to do a discretization. In addition, the continuity Equation (16) is first discretized before substituting the discrete version of (14). To approximate the problem in space, the finite difference method is used. Let the equations in discrete form be given by

$$D_h \mathbf{V}_h^{(n+1)} = 0, \qquad (19)$$

$$\frac{1}{\delta t}(\mathbf{V}_h^{(n+1)} - \mathbf{V}_h^{(n)}) + G_h P_h^{(n+1)} = \mathbf{R}_h^{(n+1)}, \qquad (20)$$

$$\mathbf{V}_h^{(n+1)}|_\Gamma = 0, \qquad (21)$$

where D_h and G_h are the discrete div and ∇ operator, respectively. Then, $\mathbf{V}_h$, P_h and $\mathbf{R}_h$ are the discrete grid functions corresponding with $\mathbf{V}$, P, and $\mathbf{R}$. After discretization of (16), the number of velocity unknowns equals the number of discrete momentum equations. The number of pressure unknowns is equal to the number of discrete continuity equations, since both are equal to the number of grid cells [23]. Our approach uses the idea of [23], but it differs in implementation.

The uniform grid Ω is introduced in the domain Ω with steps h_1 and h_2; $h_1 = 1/N_1$, $h_2 = 1/N_2$, where N_1, N_2 are the number of cells in each direction. The grid cells are positioned such that the cell faces coincide with the boundary Γ of Ω. The discretization in space of the Navier–Stokes equations is performed on MAC (Marker and Cell) [24] (staggered) grids when pressure P and velocities in two-dimensional problems are determined on three grids shifted relative to each other. So, P is located in the center of each cell, the x-component velocity u is on the middle points of vertical faces, the y-component velocity v is on the middle points of horizontal faces. For the MAC-method, the solution advanced in time by solving the momentum equation with the best current estimate of pressure distribution. Such a solution initially would not satisfy the continuity equation unless the correct pressure distribution was used. The pressure is improved by numerically solving the Poisson equation with estimated velocity field. We rewrite the equation for pressure in the following form:

$$\Delta P = \frac{d}{dx}\left(-\left(u\frac{\partial u}{\partial x} + v\frac{\partial u}{\partial y}\right) + \frac{1}{Re}\left(\frac{\partial^2 u}{\partial x^2} + \frac{\partial^2 u}{\partial y^2}\right)\right) + \qquad (22)$$

$$+ \frac{d}{dy}\left(-\left(u\frac{\partial v}{\partial x} + v\frac{\partial v}{\partial y}\right) + \frac{1}{Re}\left(\frac{\partial^2 v}{\partial x^2} + \frac{\partial^2 v}{\partial y^2}\right)\right) - \frac{\partial}{\partial t}\left(\frac{\partial u}{\partial x} + \frac{\partial v}{\partial y}\right).$$

We now introduce the grid sets and the corresponding spaces:

$$\overline{D}_1 = \{x_{ij} = ((i+1/2)h_1, jh_2) : i = 0, ..., N_1 - 1, j = 0, ..., N_2\},$$

$$\overline{D}_2 = \{x_{ij} = (ih_1, (j+1/2)h_2) : i = 0, ..., N_1, j = 0, ..., N_2 - 1\},$$

$$D_3 = \{x_{ij} = (ih_1, jh_2) : i = 1, ..., N_1 - 1, j = 1, ..., N_2 - 1\}.$$

Let $\mathbf{V}_h = V_{1,h} \times V_{2,h}$ be the linear space of vector functions defined on $\overline{D}_1 \times \overline{D}_2$ and vanishing at the corresponding grid boundaries, and P_h is the space of functions defined on D_3 and orthogonal to unity. Thus,

$$V_{1,h} = \{u_{ij} = u(x_{ij}) : x_{ij} \in \overline{D}_1, u_{0,j} = u_{N_1-1,j} = u_{i,0} = u_{i,N_2} = 0\},$$
$$V_{2,h} = \{v_{ij} = v(x_{ij}) : x_{ij} \in \overline{D}_2, v_{0,j} = v_{N_1,j} = v_{i,0} = v_{i,N_2-1} = 0\},$$
$$P_h = \{P_{ij} = P(x_{ij}) : x_{ij} \in D_3, \sum_{ij} h_1 h_2 P_{ij} = 0\}.$$

Variables are denoted by a single set of indices, despite the fact that different variables are calculated at different grid nodes. As a result, the indices i, j refer to a set of three mismatched points.

The term $\mathbf{R}^{(n+1)}$ in (15) contains the nonlinear terms. So, for treating this nonlinearity, Newton linearization around the old time level is used. For example, we want to linearize a nonlinear term $u^{(n+1)}\phi_x^{(n+1)}$, then

$$u^{(n+1)}\phi_x^{(n+1)} = u^{(n)}\phi_x^{(n+1)} + u^{(n+1)}\phi_x^{(n)} - u^{(n)}\phi_x^{(n)} + O(\delta t^2). \tag{23}$$

The expression in the right-hand side of (23) is linear in the variables at the new time level and possesses a discretization error $O(\delta t^2)$.

Let $D = \dfrac{\partial u}{\partial x} + \dfrac{\partial v}{\partial y}$ in (22) be the local dilation term, and other terms with velocity field determined from the solution of momentum equation with a provisional estimate of pressure $P' = \binom{f_1'}{f_2'}$, counter, and $D_{ij}^{(n+1)}$ be set equal to zero. That is, the correction of pressure is required to compensate for nonzero dilation at the n iterative level. The Poisson equation is then solved for the revised pressure field. The improved pressure is then used in the momentum equation for better solution at time step. If the dilation (divergence of velocity field) is not zero, the cyclic process of solving the momentum equation and Poisson equation is repeated until the velocity field is divergence free.

Thus, our computational scheme can be represented as follows:

1. Velocity field components $u' - u^{(n+1)}$ and $v' - v^{(n+1)}$ are determined by solving the implicit momentum equation with P', and for treating nonlinearity, the Newton linearization around the old time level is used.

$$\frac{u_{ij}^{(n+1)} - u_{ij}^{(n)}}{\delta t} + \left(u_{ij}^{(n)} \left(\frac{u_{ij}^{(n+1)} - u_{i-1,j}^{(n+1)}}{h_1} \right) + u_{ij}^{(n+1)} \left(\frac{u_{ij}^{(n)} - u_{i-1,j}^{(n)}}{h_1} \right) - u_{ij}^{(n)} \left(\frac{u_{ij}^{(n)} - u_{i-1,j}^{(n)}}{h_1} \right) \right) +$$

$$+ \left(v_{ij}^{(n)} \left(\frac{u_{ij}^{(n+1)} - u_{i,j-1}^{(n+1)}}{h_2} \right) + v_{ij}^{(n+1)} \left(\frac{u_{ij}^{(n)} - u_{i,j-1}^{(n)}}{h_2} \right) - v_{ij}^{(n)} \left(\frac{u_{ij}^{(n)} - u_{i,j-1}^{(n)}}{h_2} \right) \right) - \tag{24}$$

$$- \frac{1}{Re} \left(\frac{u_{i+1,j}^{(n+1)} - 2u_{ij}^{(n+1)} + u_{i-1,j}^{(n+1)}}{h_1^2} + \frac{u_{i,j+1}^{(n+1)} - 2u_{ij}^{(n+1)} + u_{i,j-1}^{(n+1)}}{h_2^2} \right) = f_1',$$

$$\frac{v_{ij}^{(n+1)} - v_{ij}^{(n)}}{\delta t} + \left(u_{ij}^{(n)} \left(\frac{v_{ij}^{(n+1)} - v_{i-1,j}^{(n+1)}}{h_1} \right) + u_{ij}^{(n+1)} \left(\frac{v_{ij}^{(n)} - v_{i-1,j}^{(n)}}{h_1} \right) - u_{ij}^{(n)} \left(\frac{v_{ij}^{(n)} - v_{i-1,j}^{(n)}}{h_1} \right) \right) +$$

$$+ \left(v_{ij}^{(n)} \left(\frac{v_{ij}^{(n+1)} - v_{i,j-1}^{(n+1)}}{h_2} \right) + v_{ij}^{(n+1)} \left(\frac{v_{ij}^{(n)} - v_{i,j-1}^{(n)}}{h_2} \right) - v_{ij}^{(n)} \left(\frac{v_{ij}^{(n)} - v_{i,j-1}^{(n)}}{h_2} \right) \right) - \tag{25}$$

$$-\frac{1}{Re}\left(\frac{v_{i+1,j}^{(n+1)} - 2v_{ij}^{(n+1)} + v_{i-1,j}^{(n+1)}}{h_1^2} + \frac{v_{i,j+1}^{(n+1)} - 2v_{ij}^{(n+1)} + v_{i,j-1}^{(n+1)}}{h_2^2}\right) = f_2',$$

$$\frac{u_{ij}^{(n+1)} - u_{i-1,j}^{(n+1)}}{h_1} + \frac{v_{ij}^{(n+1)} - v_{i,j-1}^{(n+1)}}{h_2} = 0. \qquad (26)$$

2. The Poisson equation with estimated velocity field components $u' = u^{(n+1)}$ and $v' = v^{(n+1)}$ is solved for the revised pressure field $P = P^{(n+1)}$.

$$\frac{1}{h_1^2}\left(P_{i+1,j}^{(n+1)} - 2P_{i,j}^{(n+1)} + P_{i-1,j}^{(n+1)}\right) + \frac{1}{h_2^2}\left(P_{i,j+1}^{(n+1)} - 2P_{i,j}^{(n+1)} + P_{i,j-1}^{(n+1)}\right) =$$

$$= \frac{1}{\delta t}\left(\frac{u_{ij}^{(n+1)} - u_{i-1,j}^{(n+1)}}{h_1} + \frac{v_{ij}^{(n+1)} - v_{i,j-1}^{(n+1)}}{h_2}\right) +$$

$$+ \frac{1}{h_1}\left(-u_{ij}^{(n+1)}\left(\frac{v_{ij}^{(n+1)} - v_{i-1,j}^{(n+1)}}{h_1}\right) - v_{ij}^{(n+1)}\left(\frac{v_{ij}^{(n+1)} - v_{i,j-1}^{(n+1)}}{h_2}\right)\right) +$$

$$+ \frac{1}{Reh_1}\left(\frac{u_{i+1,j}^{(n+1)} - 2u_{ij}^{(n+1)} + u_{i-1,j}^{(n+1)}}{h_1^2} + \frac{u_{i,j+1}^{(n+1)} - 2u_{ij}^{(n+1)} + u_{i,j-1}^{(n+1)}}{h_2^2}\right) +$$

$$+ \frac{1}{h_2}\left(-u_{ij}^{(n+1)}\left(\frac{v_{ij}^{(n+1)} - v_{i-1,j}^{(n+1)}}{h_1}\right) - v_{ij}^{(n+1)}\left(\frac{v_{ij}^{(n+1)} - v_{i,j-1}^{(n+1)}}{h_2}\right)\right) +$$

$$+ \frac{1}{Reh_2}\left(\frac{v_{i+1,j}^{(n+1)} - 2v_{ij}^{(n+1)} + v_{i-1,j}^{(n+1)}}{h_1^2} + \frac{v_{i,j+1}^{(n+1)} - 2v_{ij}^{(n+1)} + v_{i,j-1}^{(n+1)}}{h_2^2}\right).$$

Revised velocity field components u and v are determined by solving the implicit momentum equation with revised pressure P. Process of solving the momentum equation and Poisson equation is repeated until the velocity field is divergence free. Thus, at each time step in solving the Navier–Stokes equation, we need to solve SLAE with nonsymmetric matrices that are solved by the AMG method with HSS smoothers.

There are two coarsening approaches in the AMG: RS and PMIS algorithms. Coarsening splits initial grid on C-points and F-points—coarse and fine grid points, respectively. The RS (Ruge-Stuben) algorithm [25] is a traditional coarsening approach. The RS algorithm is based on two heuristic criteria that achieve optimal convergence and minimal computational cost. The first criterion provides the achievement of good convergence, as the effective coarsening scheme should allow to accurately interpolate a smooth error. Then, it is desirable that each F-point (Fine-grid point) has as many strongly influencing C-points (Coarse-grid points) as possible [26]. The criterion, provided minimal computational cost for different levels of V-cycle, requires that the set of C-points is the maximum subset of all F-points, to obtain more accurate interpolation, provided that no C-point is strongly dependent on another C-point (the set is maximum and independent), since such points would have increased the computational costs without providing visible benefits of interpolation [26]. In general, as the convergence is increased, the computational costs of the V-cycle decrease. Therefore, the first criterion is strictly observed and the second one is guidance. The RS algorithm has two passes. The first pass splits the full grid in C and F points; the second one ensures strict implementation of the first criterion [26]. PMIS (parallel changes independent set), the algorithm of coarsening, is based on the same principles as the RS algorithm except that a heuristic criterion is not strictly observed, i.e., F-F connections without a common C-point are permitted. Unlike the RS coarsening, the PMIS is not sequential. However, the precision may be deteriorated because an insufficient number of points reduces the accuracy of interpolation [26].

Numerical experiments have been done using the PMIS-algorithm. In Tables 1 and 2, we give the number of AMG-iterations with the HSS-based smoother on the different grids, where α is the parameter of the HSS iteration method. For comparison, we give the AMG calculations when the Gauss–Seidel method is used as the smoothing procedure. In our implementations, all iterations are started from the zero vector, and terminated when

$$\frac{\|r^{(p)}\|_2}{\|r^{(0)}\|_2} \leq 10^{-6},$$

where $r^{(p)} = b - Av^{(p)}$ is the residual vector of the SLAE (2) at the current iterate $v^{(p)}$ and $r^{(0)}$ is the initial residual. Our comparisons are done for the number of iteration steps and the elapsed CPU time (in seconds, in parentheses). The abbreviation "n.c." in Table 2 means "no convergence".

The experiments are run in MATLAB (version R2018b) with a machine precision of 10^{-16}.

From Tables 1 and 2, it follows that the AMG methods with the HSS-smoother have fast convergence speed for all tested values of the viscosity coefficient ($\nu = 10^{-1} \div 10^{-5}$) on all used grids, while the AMG with the Gauss–Seidel smoother does not converge for $\nu = 10^{-4}, 10^{-5}$ on all considered grids, and does not converge on the grids 260×260 and 520×520 nodes for all values of the viscosity coefficient. For all tests, the AMG+HSS (Algebraic multigrid method with Hermitian/Skew-Hermitian Splitting smoother) outperforms the AMG+GS (Algebraic multigrid method with Gauss–Seidel smoother) with respect to both number of iteration steps and CPU time. Moreover, the number of iteration steps and CPU time increase with increasing grid size for both methods.

From the data shown in the Table 1, an increase in the number of iterations with an increase in the mesh size follows. However, this relates to some features of the algebraic approach in MGM (more precisely, the PMIS algorithm in the AMG). The traditional (a scalable) approach in the AMG (RS algorithm) works well for problems arising from the discretization of PDEs in two spatial dimensions. For many two-dimensional problems, a solver can be obtained with the number of iterations, regardless of the size of the problem n, as well as the solution time per iteration, linearly proportional to n. For the RS algorithm, the convergence factor is separated from unity and does not depend on the size of the problem n. But when using regular AMG interpolation in combination with PMIS, AMG convergence worsens depending on the size of the problem. This results in a loss of scalability [27]. However, when traditional AMG algorithms are applied to three-dimensional (3D) problems, numerical tests show [27] that in many cases scalability is lost. However, the number of iterations may remain constant. The computational complexity and size of the stencil can increase significantly, which will lead to an increase in execution time and memory usage. In addition, the PMIS algorithm allows for natural parallelization, unlike the RS algorithm. These properties of the PMIS algorithm seem promising to us for the further study of the three-dimensional Navier–Stokes equations using parallel computing.

Table 1. Algebraic multigrid (AMG)+(HSS) Hermitian/skew-Hermitian splitting iterations with different ν.

Grid	$\nu = 10^{-1}$	$\nu = 10^{-2}$	$\nu = 10^{-3}$	$\nu = 10^{-4}$	$\nu = 10^{-5}$
60×60	25 (21.20)	26 (21.59)	29 (26.20)	30 (26.54)	21 (14.15)
120×120	40 (64.50)	45 (67.70)	54 (94.60)	40 (59.20)	37 (51.50)
180×180	54 (152.61)	50 (151.52)	64 (161.82)	49 (126.85)	35 (97.51)
260×260	85 (192.70)	93 (197.20)	82 (191.52)	83 (197.26)	58 (126.7)
520×520	90 (282.51)	97 (290.58)	90 (286.26)	92 (286.85)	85 (252.21)

Table 2. Algebraic multigrid (AMG)+(GS) Gauss–Seidel iterations with different v.

Grid	$v = 10^{-1}$	$v = 10^{-2}$	$v = 10^{-3}$	$v = 10^{-4}$	$v = 10^{-5}$
60×60	26 (32.57)	46 (42.51)	54 (114.85)	n.c.	n.c.
120×120	57 (83.82)	64 (122.61)	83 (160.50)	n.c.	n.c.
180×180	59 (162.36)	82 (185.38)	85 (192.20)	n.c.	n.c.
260×260	n.c.	n.c.	n.c.	n.c.	n.c.
520×520	n.c.	n.c.	n.c.	n.c.	n.c.

Table 3 shows the number of iteration steps and CPU time of the AMG+HSS method depending on the value α, when $v = 10^{-5}$. For the AMG+HSS method, the optimal (experimental) parameter value that reduces the number of iterations depends on the size of the grid. As the grid size increases, the value of α, which provides the best convergence, decreases. Numerical experiments showed that for parameter values less than 0.2, the AMG+HSS method diverges.

Table 3. (AMG)+(HSS) iterations with different α, $v = 10^{-5}$.

Grid	$\alpha = 0.2$	$\alpha = 0.3$	$\alpha = 0.4$	$\alpha = 0.6$	$\alpha = 0.8$	$\alpha = 0.9$	$\alpha = 1.0$
60×60	29 (26.84)	24 (21.51)	21 (14.15)	42 (40.61)	54 (58.86)	56 (68.22)	65 (84.20)
120×120	40 (61.50)	39 (64.67)	37 (51.50)	45 (67.52)	56 (94.60)	57 (94.20)	82 (162.85)
180×180	52 (114.2)	35 (97.51)	42 (129.20)	67 (14.82)	84 (165.84)	86 (175.21)	91 (196.21)
260×260	58 (126.7)	65(171.58)	65 (187.21)	82 (192.64)	84 (194.54)	91 (194.60)	95 (197.22)
520×520	82 (251.26)	84(251.84)	85 (252.21)	92 (260.52)	93 (262.42)	94 (282.52)	97 (290.21)

Thus, the numerical experiments have showed that the HSS-based smoothers can be effectively used for the AMG, in which the stage of coarse-grid correction can be considered as a kind of accelerating procedure of the HSS methods.

5. Conclusions

In our previous theoretical and numerical studies of the MGM with the STS-based smoothers, the stationary (and nonstationary) linear diffusion–convection equation with dominant convection was considered as a test problem [13,14]. All theoretical results and calculations were performed using geometric MGM. Here, we first use the HSS-method as the smoother in the algebraic MGM for solving the unsteady Navier–Stokes equations. It is supposed to further prove the theoretically smoothing properties of the HSS iteration methods and to prove the convergence of the MGM with the corresponding smoothers. In addition, theoretical and numerical results should be obtained for the MGM with the STS-based smoothers for the Navier–Stokes problem. The PMIS algorithm was not chosen by us by chance. Preliminary testing of it on this model problem showed its robustness. In addition, the PMIS algorithm allows for natural parallelization, unlike the RS algorithm. These properties of the PMIS algorithm seem promising to us for the further study of the three-dimensional Navier–Stokes equations using parallel computing.

Author Contributions: Conceptualization, G.M.; methodology, G.M., T.M.; validation, E.A.; writing—original draft preparation, T.M., Z.-Q.W.; writing—review and editing, T.M.; software, V.B. All authors have read and agreed to the published version of the manuscript.

Funding: This research was funded by RFBR, grant N19-51-53013 GFENa, and Ministry of Science and Higher Education of the Russian Federation (basic part, project N1.5169.2017/8.9).

Conflicts of Interest: The authors declare no conflict of interest. The funders had no role in the design of the study; in the collection, analyses, or interpretation of data; in the writing of the manuscript, or in the decision to publish the results.

Abbreviations

The following abbreviations are used in this manuscript:

MGM	Multigrid Method
SLAE	Systems of Linear Algebraic Equations
AMG	Algebraic Multigrid
HSS	Hermitian/Skew-Hermitian Splitting
STS	Skew-Hermitian Triangular Splitting
PDE	Partial Differential Equations
CFD	Computational Fluid Dynamics

References

1. Bai, Z.-Z.; Golub, G.H.; Ng, M.K. Hermitian and skew-Hermitian splitting methods for non-Hermitian positive definite linear systems. *SIAM J. Matrix Anal. Appl.* **2003**, *24*, 603–626. [CrossRef]
2. Bai, Z.-Z. Splitting iteration methods for non-Hermitian positive definite systems of linear equations. *Hokkaido Math. J.* **2007**, *36*, 801–814. [CrossRef]
3. Bai, Z.-Z.; Golub, G.H.; Ng, M.K. On inexact Hermitian and skew-Hermitian splitting methods for non-Hermitian positive definite linear systems. *Linear Algebra Appl.* **2008**, *428*, 413–440. [CrossRef]
4. Bai, Z.-Z.; Golub, G.H.; Pan, J.-Y. Preconditioned Hermitian and skew-Hermitian splitting methods for non-Hermitian positive semidefinite linear systems. *Numer. Math.* **2004**, *98*, 1–32. [CrossRef]
5. Bai, Z.-Z.; Golub, G.H.; Lu, L.-Z.; Yin, J.-F. Block triangular and skew-Hermitian splitting methods for positive definite linear systems. *SIAM J. Sci. Comput.* **2005**, *26*, 844–863. [CrossRef]
6. Krukier, L.A. Implicit Difference Schemes and an Iterative Method for Their Solution for One Class of Quasilinear Systems of Equations. *Izv. Vuzov. Math.* **1979**, *7*, 41–52, (In Russian).
7. Krukier, L.A. Convergence acceleration of triangular iterative methods based on the skew-symmetric part of the matrix. *Appl. Numer. Math.* **1999**, *30*, 281–290. [CrossRef]
8. Krukier, L.A.; Chikina, L.G.; Belokon, T.V. Triangular skew-symmetric iterative solvers for strongly nonsymmetric positive real linear system of equations. *Appl. Numer. Math.* **2002**, *41*, 89–105. [CrossRef]
9. Wang, L.; Bai, Z.-Z. Skew-Hermitian triangular splitting iteration methods for non-Hermitian positive definite linear systems of strong skew-Hermitian parts. *BIT Numer. Math.* **2004**, *44*, 363–386. [CrossRef]
10. Botchev, M.A.; Krukier, L.A. On an iterative solution of strongly nonsymmetric systems of linear algebraic equations. *J. Comput. Math. Math. Phys.* **1997**, *37*, 1283–1293. (In Russian).
11. Krukier, L.A.; Martinova, T.S.; Bai, Z.-Z. Product-Type Skew-Hermitian Triangular Splitting Iteration Methods for Strongly Non-Hermitian Positive Definite Linear Systems. *J. Comput. Appl. Math.* **2009**, *232*, 3–16. [CrossRef]
12. Bai, Z.-Z.; Krukier, L.A.; Martinova, T.S. Two-step iterative methods for solving the stationary convection-diffusion equation with a small parameter for the highest derivative on a uniform grid. *J. Comput. Math. Math. Phys.* **2006**, *46*, 295–306. (In Russian) [CrossRef]
13. Muratova, G.V.; Andreeva, E.M. Multigrid method for solving convection-diffusion problems with dominant convection. *J. Comput. Appl. Math.* **2009**, *226*, 77–83. [CrossRef]
14. Muratova, G.V.; Krukier, L.A.; Andreeva, E.M. Fourier analysis of multigrid method with triangular skew-symmetric smoothers. *Commun. Appl. Math. Comput.* **2013**, *27*, 355–362.
15. Li, S.; Huang, Z. Convergence analysis of HSS-multigrid methods for second-order nonselfadjoint elliptic problems. *BIT Numer. Math.* **2013**, *53*, 987–1012. [CrossRef]
16. Fedorenko, R.P. The relaxation method for solving difference elliptic equations. *J. Comput. Math. Math. Phys.* **1961**, *1*, 922–927. (In Russian) [CrossRef]
17. Brandt, A. Multi-level adaptive solutions to boundary-value problems. *Math. Comput.* **1977**, *31*, 333–390. [CrossRef]
18. Hackbusch, W. Convergence of multi-grid iterations applied to difference equations. *Math. Comput.* **1980**, *34*, 425–440. [CrossRef]
19. Briggs, W.; Henson, V.; McCormick, S. Algebraic Multigrid (AMG). In *A Multigrid Tutorial*, 2nd ed.; Poulson, D., Briggeman, L., Eds.; SIAM Publications: Philadelphia, PA, USA, 2000; pp. 137–159.
20. Falgout, R. An introduction to algebraic multigrid. *Comput. Sci. Eng.* **2006**, *8*, 24–33. [CrossRef]
21. Hackbusch, W. Introductory Model Problem. In *Multigrid Method and Application*; Graham, R.L., Jolla, L., Eds.; Springer: Berlin/Heidelberg, Germany, 1985; pp. 17–39.

22. Brandt, A.; Yavneh, I. Accelerated multigrid convergence and highReynolds recirculating flows. *SIAM J. Sci. Comput.* **1993**, *14*, 151–164. [CrossRef]
23. Veldman, A.E.P. "Missing" boundary conditions? Discretize first, substitute next, and combine later. *SIAM J. Sci. Stat. Comput.* **1990**, *11*, 82–91. [CrossRef]
24. McKee, S.; Tome, M.F.; Ferreira, V.G.; Cuminato, J.A.; Castelo, A.; Sousa, F.S.; Mangiavacchi, N. The MAC method. *Comput. Fluids* **2008**, *37*, 907–930. [CrossRef]
25. Stuben, K. Algebraic multigrid (amg): Experiences and comparisons. *Appl. Math. Comput.* **1983**, *13*, 419–451. [CrossRef]
26. Yang, U.M. Parallel Algebraic Multigrid Methods—High Performance Preconditioners. In *Numerical Solution of Partial Differential Equations on Parallel Computers*; Lecture Notes in Computational Science and Engineering; Bruaset, A.M., Tveito, A., Eds.; Springer: Berlin/Heidelberg, Germany, 2006; Volume 51, pp. 209–236.
27. Sterck, H.; Yang, U.M.; Heys, J. Reducing Complexity in Parallel Algebraic Multigrid Preconditioners. *SIAM J. Matrix Anal. Appl.* **2006**, *27*, 1019–1039. [CrossRef]

 © 2020 by the authors. Licensee MDPI, Basel, Switzerland. This article is an open access article distributed under the terms and conditions of the Creative Commons Attribution (CC BY) license (http://creativecommons.org/licenses/by/4.0/).

Article

Discrete Symmetry Group Approach for Numerical Solution of the Heat Equation

Khudija Bibi [1,2,*] **and Tooba Feroze** [1]

[1] Department of Mathematics, School of Natural Sciences, National University of Science and Technology Islamabad 44000, Pakistan; tferoze@sns.nust.edu.pk
[2] Department of Mathematics, Faculty of Basic ans Applied Sciences, International Islamic University Islamabad 44000, Pakistan
* Correspondence: khudija.bibi@iiu.edu.pk

Received: 17 January 2020; Accepted: 18 February 2020; Published: 2 March 2020

Abstract: In this article, an invariantized finite difference scheme to find the solution of the heat equation, is developed. The scheme is based on a discrete symmetry transformation. A comparison of the results obtained by the proposed scheme and the Crank Nicolson method is carried out with reference to the exact solutions. It is found that the proposed invariantized scheme for the heat equation improves the efficiency and accuracy of the existing Crank Nicolson method.

Keywords: lie symmetries; invariantized difference scheme; numerical solutions

1. Introduction

Lie's theory of symmetry groups for differential equations was initiated and utilized for obtaining solutions or reductions of the differential equations [1]. Lie constructed a highly algorithmic technique for the solutions of differential equations. Lie established that several techniques to find the solutions of differential equations can be described and deduced by considering the Lie group analysis of differential equations [2]. Thus, Lie symmetry methodology has a great significance in the theory and applications of differential equations. It is broadly applied by several researchers to solve difficult nonlinear problems. Lie studied the groups of continuous transformations. These transformations (symmetries) can be described by the infinitesimal generators.

The symmetries, which are not continuous are called discrete symmetries. Discrete symmetries have several applications in differential equations, e.g., they are used to simplify the numerical scheme and to find the new exact solutions from the known solutions [3]. The nature of bifurcations in nonlinear dynamical systems are also obtained by using discrete symmetry groups [4]. The procedure to find the discrete symmetries of the differential equations is discussed in [5–8].

Nowadays, it is a challenging area of research to solve the dynamical equations, as various phenomena in nature are modeled in dynamical systems. Many researchers have considered dynamical systems, e.g., Martinez, Yu Zhang, and Timothy Gordon studied the uses of the control scheme in the classical dynamical systems theory to predict driver behavior and vehicle trajectories [9]. Martinez and Timothy also discussed the uses of machine learning for the systematic understanding of human control behaviors in driving [10]. Lie group theory has became a universal tool for the analysis of dynamical equations. Lie symmetry analysis provides an effective way to solve the partial differential equations.

But recently, interest is rising on the applications of Lie group analysis in the partial differential equations for their numerical solutions. Some work has been dedicated to building those numerical schemes that preserve the symmetries of the given differential equations. Invariantized finite difference schemes by using the idea of moving frame were constructed by Kim [11] and Olver [12]. The technique of discretization that preserves some continuous symmetries of the original differential equation was also studied in [13–17]. In [18], the exact solutions of Fisher's type equation with the help of Lie symmetries, which are continuous symmetries in nature, are studied. Since most of the partial differential equations have some geometrical properties and some discrete symmetry groups correspond to these geometrical features of partial differential equations [8], the new invaiantized finite difference methods constructed by using these discrete symmetry groups may show better performances than the other finite difference methods.

In this article, first, it is shown that the Crank Nicolson scheme of a diffusion (heat) equation is invariant under a group of the discrete symmetry transformation. Here, a modification is proposed for the invariantization of the Crank Nicolson method given by Kim et al. [19]. This modification is proposed with the help of the composition of discrete and continuous symmetry transformations of the heat equation. It is also shown that the proposed invariantized scheme gives better results, as compared to the other classical finite difference methods.

2. Heat Equation

The homogeneous heat (diffusion) [20] equation

$$\frac{\partial w}{\partial t} - D\frac{\partial^2 w}{\partial x^2} = 0, \tag{1}$$

plays a vital role in the study of heat conduction and other diffusion processes, in which a thin metal rod of length L, whose sides are insulated, is considered. The temperature of the bar at the point x and at the time t is represented by $w(x, t)$. The parameter D is called the thermal conductivity and it depends only upon the material from which the rod is made. For simplicity, we take the parameter $D = 1$ then the Equation (1) becomes

$$\frac{\partial w}{\partial t} - \frac{\partial^2 w}{\partial x^2} = 0. \tag{2}$$

2.1. Continuous Symmetry Groups

Equation (2) is a linear and homogenous partial differential equation with one dependent variable w and two independent variables x and t. The point transformation of Equation (2) defines the local diffeomorphism

$$\Gamma : (x, t, w) \mapsto (\hat{x}(x, t, w), \hat{t}(x, t, w), \hat{w}(x, t, w)).$$

This transformation maps any surface $w = f(x, t)$ to the following surface

$$\hat{x} = \hat{x}(x, t, f(x, t)),$$

$$\hat{t} = \hat{t}(x, t, f(x, t)),$$

$$\hat{w} = \hat{w}(x, t, f(x, t)).$$

The heat Equation (2) has infinite dimensional Lie algebra. The infinitesimal generators and the continuous symmetry groups of the Equation (2) [21] are presented in Table 1.

Table 1. Continuous symmetry transformations of (2).

	Generators	Symmetry Transformations
1	$\frac{\partial}{\partial x}$	Space translation: $(x, t, w) \mapsto (x + \epsilon, t, w)$
2	$\frac{\partial}{\partial t}$	Time translation: $(x, t, w) \mapsto (x, t + \epsilon, w)$
3	$w\frac{\partial}{\partial w}$	Scale Transformation: $(x, t, w) \mapsto (x, t, e^\epsilon w)$
4	$x\frac{\partial}{\partial x} + 2t\frac{\partial}{\partial t}$	Scale Transformation: $(x, t, w) \mapsto (e^\epsilon x, e^{2\epsilon} t, w)$
5	$2t\frac{\partial}{\partial x} - xw\frac{\partial}{\partial w}$	Galilean boost: $(x, t, w) \mapsto (x + 2\epsilon t, t, -xw\epsilon + w)$
6	$4xt\frac{\partial}{\partial x} + 4t^2\frac{\partial}{\partial t} - (x^2 + 2t)w\frac{\partial}{\partial w}$	Projection: $(x, t, w) \mapsto \left(\frac{x}{1-4\epsilon t}, \frac{t}{1-4\epsilon t}, w\sqrt{1-4\epsilon t}\exp(\frac{-\epsilon x^2}{1-4\epsilon t})\right)$
7	$q(x, t)\frac{\partial}{\partial w}$	

Due to the arbitrariness of the function appearing in the last generator of Table 1, Lie algebra of Equation (2) is infinite dimensional. Each of the groups given in Table 1 has the property of mapping solutions of heat Equation (2) to the other solutions. For example, consider the explicit form of the projection

$$(x, t, w) \mapsto \left(\frac{x}{1 - 4\epsilon t}, \frac{t}{1 - 4\epsilon t}, w\sqrt{1 - 4\epsilon t} \exp\left(\frac{-\epsilon x^2}{1 - 4\epsilon t}\right) \right),$$

where ϵ is the group parameter. Computing the induced action on graphs of functions, we conclude that if $w = f(x, t)$ is any solution to heat Equation (2), so is

$$w = \frac{1}{\sqrt{1 - 4\epsilon t}} \exp\left(\frac{\epsilon x^2}{1 - 4\epsilon t}\right) f\left(\frac{x}{1 - 4\epsilon t}, \frac{t}{1 - 4\epsilon t}\right).$$

2.2. Discrete Symmetry Group

Hydon introduced a technique [7] by which all discrete point symmetries of the partial differential equations can be found on basis of the results [6] that each continuous symmetry generator of a Lie algebra ℓ of a differential equation brings an automorphism that preserve the following commutator relation

$$[X_l, X_k] = c_{lk}^m X_m.$$

Hydon's method categorizes and factor out all those automorphisms of a Lie algebra ℓ that are equivalent under the action of a Lie symmetry in a Lie group that is generated by the Lie algebra ℓ and provide the most general realization of these automorphisms as point transformations. Finally, by using these point transformations, an entire list of discrete point symmetries of a partial differential equation, is obtained.

The discrete symmetry group of Equation (2) has already been obtained in [8], which is isomorphic to Z_4 = {group of residues modulo 4} and is generated by

$$(x, t, w) \to \left(\frac{x}{2t}, \frac{-1}{4t}, \sqrt{2\iota t} \exp\left(\frac{x^2}{4t}\right) w \right), \tag{3}$$

where $\iota = \sqrt{-1}$.

3. Finite Difference Schemes for the Heat Equation

Some finite difference schemes are available in the literature, which help us to find the numerical solutions of the partial differential equations. In this section, the backward difference scheme, forward difference scheme, and the Crank Nicolson method for the heat equation along their stability conditions, are given [22].

3.1. Forward Difference Scheme

The forward difference scheme for heat Equation (2) is given by

$$w_{n,j+1} = \alpha w_{n+1,j} + (1 - 2\alpha) w_{n,j} + \alpha w_{n-1,j},$$

where $\alpha = \frac{k}{h^2}$. Here h is the x-axis step size and k is the time step size for the grid points (x_i, t_j), where $x_i = ih$, $t_j = jk$ for non-negative integers i, j. This scheme is Forward in Time and Centered in Space (FTCS). This method is explicit and converges to the solution for $0 < \alpha \leq \frac{1}{2}$, so is conditionally stable [22].

3.2. Backward Difference Scheme

The backward difference scheme for Equation (2) is given by

$$w_{n,j-1} = -\alpha w_{n+1,j} + (1+2\alpha)w_{n,j} - \alpha w_{n-1,j}.$$

This scheme is Backward in Time and Centered in Space (BTCS). This is an implicit and unconditionally stable scheme [22].

3.3. Crank Nicolson Method

The Crank Nicolson method (CNM) for Equation (2) is

$$\alpha w_{n-1,j} + 2(1-\alpha)w_{n,j} + \alpha w_{n+1,j} = -\alpha w_{n-1,j+1}$$

$$+2(1+\alpha)w_{n,j+1} - \alpha w_{n+1,j+1}. \quad (4)$$

The Crank Nicolson method is also implicit and unconditionally stable [22]. It has significant advantages for the time-accurate solutions. The temporal truncation error of CNM is $O(\triangle t^2)$, whereas the truncation error of FTCS and BTCS is $O(\triangle t)$.

3.4. Invariantization of the Crank Nicolson Method under the Discrete Symmetry Transformation

It is observed that among the mentioned finite difference schemes for the heat equation, the Crank Nicolson method gives the more accurate results [23], so we are interested in constructing an invariantization of the Crank Nicolson method. In the present subsection, we show that the Crank Nicolson method is invariant under the discrete symmetry transformation (3).

Let $w_{n,j} = w(n\triangle x, j\triangle t)$ be an approximation of $w(x,t)$ at the mesh point (x_n, t_j). Now, by using the discrete symmetry group of heat equation given in (3), we have the following transformation

$$w = \sqrt{2it}\exp\left(\frac{x^2}{4t}\right)w. \quad (5)$$

By using (5), the Crank Nicolson method transformed to

$$\alpha w_{n-1,j}\sqrt{2ij\triangle t}\exp\left\{(n-1)^2\frac{(\triangle x)^2}{4j\triangle t}\right\} + 2(1-\alpha)w_{n,j}\sqrt{2ij\triangle t}\exp\left\{\frac{(n\triangle x)^2}{4j\triangle t}\right\} +$$

$$\alpha w_{n+1,j}\sqrt{2ij\triangle t}\exp\left\{(n+1)^2\frac{(\triangle x)^2}{4j\triangle t}\right\} = -\alpha w_{n-1,j+1}\sqrt{2i(j+1)\triangle t}\exp\left\{(n-1)^2\frac{(\triangle x)^2}{4(j+1)\triangle t}\right\}$$

$$+2(1+\alpha)w_{n,j+1}\sqrt{2i(j+1)\triangle t}\exp\left\{\frac{(n\triangle x)^2}{4(j+1)\triangle t}\right\} -$$

$$\alpha w_{n+1,j+1}\sqrt{2i(j+1)\triangle t}\exp\left\{(n+1)^2\frac{(\triangle x)^2}{4(j+1)\triangle t}\right\}. \quad (6)$$

By considering the following transformation

$$w_{N,J} = \sqrt{J}\exp\left(\frac{N^2}{4J}\right)w_{n,j},$$

with

$$J = j\triangle t \text{ and } N = n\triangle x,$$

(6) can be written as

$$\alpha w_{N-1,J} + 2(1-\alpha)w_{N,J} + \alpha w_{N+1,J} =$$

$$-\alpha w_{N-1,J+1} + 2(1+\alpha)w_{N,J+1} - \alpha w_{N+1,J+1}. \quad (7)$$

The finite difference approximation to heat Equation (2) obtained in (7) is similar to the Crank Nicolson method for heat Equation (4). Thus, the Crank Nicolson method remains invariant under the discrete symmetry transformation (3). The consequence of the result obtained in Section 3 can be written in the form of the following theorem.

Theorem 1. *The Crank Nicolson method for heat equation given in (4) is invariant under the only discrete symmetry transformation (3) of the heat Equation (2).*

4. Discrete Symmetry Numerical Scheme For the Heat Equation

Most of the finite difference methods including the Crank Nicolson method are invariant under time, space translations, and scale transformation [24]. It is also proved in the previous literature that the Crank Nicolson method for the heat Equation (2) is invariant under the transformation of the discrete symmetry. Notice that the Galilean boost and projection group are the transformations under which the Crank Nicolson method is not invariant. However, the Crank Nicolson method invariantized by Galilean boost, projection transformations or composition of these transformations, becomes unstable (i.e., does not converge to the exact solution). Nevertheless, by taking the composition of the discrete symmetry group and the projective symmetry group, which is a continuous symmetry group of the heat equation, invariantized Crank Nicolson method converges to the exact solution and gives the more adequate results as compare to other existing finite difference schemes of Equation (2).

In the present section, we construct an invariantization of the Crank Nicolson method for Equation (2) by using the composition of discrete and continuous symmetry groups. The construction of this method is based on the composition of the variable w of these two groups. We deal with the following transformation to construct the new scheme:

$$w = \sqrt{2\iota t} \exp\left\{\frac{x^2}{4(1-4\epsilon t)}\left(\frac{1}{t} - 4\epsilon\right)\right\}w,$$

where $\iota = \sqrt{-1}$. By transforming the variable w in the Crank Nicolson method (4) with the above transformation for Equation (2), we get:

$$\alpha\sqrt{t_j}\, w_{n-1,j} \exp\left\{\frac{(x_{n-1})^2}{4(1-4\epsilon t_j)}\left(\frac{1}{t_j} - 4\epsilon\right)\right\}$$

$$+(2-2\alpha)\sqrt{t_j}\, w_{n,j} \exp\left\{\frac{(x_n)^2}{4(1-4\epsilon t_j)}\left(\frac{1}{t_j} - 4\epsilon\right)\right\}$$

$$+\alpha\sqrt{t_j}\, w_{n+1,j} \exp\left\{\frac{(x_{n+1})^2}{4(1-4\epsilon t_j)}\left(\frac{1}{t_j} - 4\epsilon\right)\right\}$$

$$= -\alpha\sqrt{t_{j+1}}\, w_{n-1,j+1} \exp\left\{\frac{(x_{n-1})^2}{4(1-4\epsilon t_{j+1})}\left(\frac{1}{t_{j+1}} - 4\epsilon\right)\right\}$$

$$+(2+2\alpha)\sqrt{t_{j+1}}\, w_{n,j+1} \exp\left\{\frac{(x_n)^2}{4(1-4\epsilon t_{j+1})}\left(\frac{1}{t_{j+1}} - 4\epsilon\right)\right\}$$

$$-\alpha\sqrt{t_{j+1}}\, w_{n+1,j+1} \exp\left\{\frac{(x_{n+1})^2}{4(1-4\epsilon t_{j+1})}\left(\frac{1}{t_{j+1}} - 4\epsilon\right)\right\},$$

which can be further simplified as

$$\sqrt{t_j}\left[\alpha w_{n-1,j} \exp\left\{\frac{(x_{n-1})^2}{4(1-4\epsilon t_j)}\left(\frac{1}{t_j} - 4\epsilon\right)\right\}\right.$$

$$+(2-2\alpha)w_{n,j}\exp\left\{\frac{(x_n)^2}{4(1-4\epsilon t_j)}\left(\frac{1}{t_j}-4\epsilon\right)\right\}+$$

$$\alpha w_{n+1,j}\exp\left\{\frac{(x_{n+1})^2}{4(1-4\epsilon t_j)}\left(\frac{1}{t_j}-4\epsilon\right)\right\}\right] =$$

$$\sqrt{t_{j+1}}\left[-\alpha w_{n-1,j+1}\exp\left\{\frac{(x_{n-1})^2}{4(1-4\epsilon t_{j+1})}\left(\frac{1}{t_{j+1}}-4\epsilon\right)\right\}\right.$$

$$+(2+2\alpha)w_{n,j+1}\exp\left\{\frac{(x_n)^2}{4(1-4\epsilon t_{j+1})}\left(\frac{1}{t_{j+1}}-4\epsilon\right)\right\}$$

$$\left.-\alpha w_{n+1,j+1}\exp\left\{\frac{(x_{n+1})^2}{4(1-4\epsilon t_{j+1})}\left(\frac{1}{t_{j+1}}-4\epsilon\right)\right\}\right].$$

Since ϵ is a continuous parameter and we can choose the optimal value of ϵ for which the proposed method gives better performance than the Crank Nicolson method. We choose $\epsilon = c_i$, where $0 \leq c_i \leq 1$ and i is a positive integer, so the above equation takes the form:

$$\sqrt{t_j}\left[\alpha w_{n-1,j}\exp\left\{\frac{(x_{n-1})^2}{4(1-4\epsilon t_j)}\left(\frac{1}{t_j}-4c_i\right)\right\}\right.$$

$$+(2-2\alpha)w_{n,j}\exp\left\{\frac{(x_n)^2}{4(1-4\epsilon t_j)}\left(\frac{1}{t_j}-4c_i\right)\right\}+$$

$$\alpha w_{n+1,j}\exp\left\{\frac{(x_{n+1})^2}{4(1-4\epsilon t_j)}\left(\frac{1}{t_j}-4c_i\right)\right\}\right] =$$

$$\sqrt{t_{j+1}}\left[-\alpha w_{n-1,j+1}\exp\left\{\frac{(x_{n-1})^2}{4(1-4\epsilon t_{j+1})}\left(\frac{1}{t_{j+1}}-4c_i\right)\right\}\right.$$

$$+(2+2\alpha)w_{n,j+1}\exp\left\{\frac{(x_n)^2}{4(1-4\epsilon t_{j+1})}\left(\frac{1}{t_{j+1}}-4c_i\right)\right\}$$

$$\left.-\alpha w_{n+1,j+1}\exp\left\{\frac{(x_{n+1})^2}{4(1-4\epsilon t_{j+1})}(\frac{1}{t_{j+1}}-4c_i\right\}\right].$$

The final form of the method is

$$A\Big(\alpha w_{n-1,j}B_1 + (2-2\alpha)w_{n,j}B_2 + \alpha w_{n+1,j}B_3\Big)$$

$$= C\Big(-\alpha w_{n-1,j+1}D_1 + (2+2\alpha)w_{n,j+1}D_2 - \alpha w_{n+1,j+1}D_3\Big), \qquad (8)$$

where

$$A = \sqrt{t_j},$$

$$B_1 = \exp\left\{\frac{(x_{n-1})^2}{4(1-4\epsilon t_j)}\left(\frac{1}{t_j}-4c_i\right)\right\},$$

$$B_2 = \exp\left\{\frac{(x_n)^2}{4(1-4\epsilon t_j)}\left(\frac{1}{t_j}-4c_i\right)\right\},$$

$$B_3 = \exp\left\{\frac{(x_{n+1})^2}{4(1-4\epsilon t_j)}\left(\frac{1}{t_j}-4c_i\right)\right\},$$

and

$$C = \sqrt{t_{j+1}},$$

$$D_1 = \exp\left\{\frac{(x_{n-1})^2}{4(1-4\epsilon t_{j+1})}\left(\frac{1}{t_{j+1}} - 4c_i\right)\right\},$$

$$D_2 = \exp\left\{\frac{(x_n)^2}{4(1-4\epsilon t_{j+1})}\left(\frac{1}{t_{j+1}} - 4c_i\right)\right\},$$

$$D_3 = \exp\left\{\frac{(x_{n+1})^2}{4(1-4\epsilon t_{j+1})}\left(\frac{1}{t_{j+1}} - 4c_i\right)\right\}.$$

Since the Crank Nicolson method for the heat equation is unconditionally stable [22] and the discritized Crank Nicolson method (DCNM) for heat Equation (2) provided in (8) is the invariantization of the Crank Nicolson method, so this method also preserves the unconditional stability condition and therefore converges to the exact solution without having any condition on α.

5. Solutions of the Heat Equation

In this section, we find the analytic solution of heat Equation (2) and compare it with the numerical solutions calculated by using the CNM and the proposed method DCNM.

5.1. Analytic Solution

We consider one dimensional homogeneous heat Equation (2) with the following boundary conditions

$$w(x,0) = g(x), \ 0 \le x \le L, \tag{9}$$

$$w(0,t) = f_1(t), \ 0 \le t \le T,$$

$$w(1,t) = f_2(t),$$

where $g(x)$, $f_1(t)$ and $f_2(t)$ are two times continuously differentiable functions on $x \in [0, L]$, L is the length of the rod and T is the maximum time.

Heat Equation (2) is reduced to an ordinary differential equation by using the following similarity variable

$$\zeta = t, \ V = \frac{w}{g(x)},$$

where $g(x)$ is obtained from the initial condition (5.1) with $g(x) \ne 0$. An exact solution of the system given by the Equations (2) and (9) for particular values of arbitrary functions $g(x)$, $f_1(t)$, and $f_2(t)$ is given in the following example.

Example 1. *Our aim is to find the solution of (2) with the following initial and boundary conditions*

$$w(x,0) = \sin\pi x, \ 0 \le x \le 1,$$

$$w(0,t) = 0, \ 0 \le t \le 1,$$

$$w(1,t) = 0.$$

Using the similarity transformation

$$\zeta = t, \ V = \frac{w}{\sin\pi x},$$

the above problem is reduced to the following ODE

$$V_\zeta + \pi^2 V = 0,$$

with the solution $V = e^{-\pi^2 t}$.

Hence $w(x,t) = e^{-\pi^2 t}\sin(\pi x)$ is the exact solution of the above boundary value problem.

5.2. Numerical Solutions Using CNM and the Proposed Method DCNM

In this subsection, the performance of the proposed method DCNM is investigated by applying it to Example 1. The efficiency of the present method DCNM is shown by calculating the absolute errors, root mean square errors L_2 and maximum errors L_∞. These errors are computed by the following formulas [18]:

$$\text{Absolute error} = |e_i|,$$

$$L_2 = \left(\sum_{i=1}^{n} e_i^2 \right)^{1/2},$$

$$L_\infty = \max_{1 \leq i \leq n} |e_i|,$$

with $e_i = (w_i - W_i)$, where w_i are numerical and W_i are the exact solutions.

To check the computational accuracy of the proposed method (DCNM), we reconsider Example 1 given in the previous subsection, which has the analytic solution $w(x,t) = \exp(-\pi^2 t)\sin(\pi x)$ [22]. We compare our results of the numerical solutions (DCNM) with the exact solutions and the solutions that are obtained by the Crank Nicolson method for the x-axis step size $h = 0.1$, a time step size $k = 0.01$ and for the time $T = 0.5$. The comparisons of the numerical solutions obtained by DCNM with the solutions calculated by FTCS, CNM, and the exact solutions are presented in Table 2, where Table 3 is showing the absolute errors of the solutions calculated by FTCS, CNM and the DCNM given in (8) for the different x-axis step sizes. Table 4 shows the values of $w(x,t)$ for fixed $h = 0.1$ and for different values of k, where Tables 5 shows the values of $w(x,t)$ for fixed $k = 0.02$ and for different values of h.

Table 2. The values of $w(x,t)$ for different x.

x_i	FTCS	CNM	DCNM	Exact Solutions
0.0	0.000000	0.000000	0.0000000000	0.000000
0.2	0.000892	0.004517	0.0042956284	0.004227
0.4	0.001444	0.007309	0.00695047277	0.006840
0.6	0.001444	0.007309	0.00695047277	0.006840
0.8	0.000892	0.004517	0.0042956284	0.004227
1.0	0.000000	0.000000	0.0000000000	0.000000

Table 3. Absolute errors for different values of x.

x_i	FTCS	CNM	DCNM
0.0	0.000000	0.000000	0.0000000000
0.2	0.003335	2.9×10^{-4}	6.8345438×10^{-5}
0.4	0.005396	4.69×10^{-4}	$1.10585242 \times 10^{-4}$
0.6	0.005396	4.69×10^{-4}	$1.10585242 \times 10^{-4}$
0.8	0.003335	2.9×10^{-4}	6.8345438×10^{-5}
1.0	0.000000	0.000000	0.000000000

Table 4. $w(x,t)$ for fixed $h = 0.1$ and for different values of k.

x_i	$k = 0.03$	$k = 0.02$	$k = 0.01$
0.0	0.0000	0.0000	0.0000
0.2	0.0053	0.0045	0.0042
0.4	0.0085	0.0076	0.0069
0.6	0.0085	0.0076	0.0069
0.8	0.0053	0.0045	0.0042
1.0	0.0000	0.0000	0.0000

Table 5. $w(x,t)$ for fixed $k = 0.02$ and for different values of h.

x_i	$h = 0.3$	$h = 0.2$	$h = 0.1$
0.0	0.0000	0.0000	0.0000
0.2	0.0050	0.0047	0.0045
0.4	0.0081	0.0078	0.0076
0.6	0.0081	0.0078	0.0076
0.8	0.0050	0.0047	0.0045
1.0	0.0000	0.0000	0.0000

Table 6 presents root mean square errors L_2 and the maximum errors L_∞ for DCNM and CNM of Example 1 for the different values of t.

Table 6. L_2 and L_∞ errors for different values of t.

t	L_2 DCNM	L_∞ DCNM	L_2 CNM	L_∞ CNM
0.1	9.6×10^{-3}	4.3×10^{-3}	4.86×10^{-3}	6.79×10^{-3}
0.3	4.0×10^{-4}	1.8×10^{-4}	8.87×10^{-5}	3.76×10^{-4}
0.5	9.0830×10^{-4}	4.0812×10^{-4}	1.73×10^{-3}	2.44×10^{-4}
0.7	1.8024×10^{-4}	1.0097×10^{-4}	2.04×10^{-4}	3.17×10^{-4}
0.9	1.0224×10^{-4}	6.1223×10^{-5}	2.14×10^{-3}	3.14×10^{-3}
1	1.1556×10^{-4}	5.0808×10^{-5}	2.15×10^{-3}	3.32×10^{-3}

Figure 1 shows the comparisons between the numerical solutions, obtained by the DCNM given in (8), the Crank Nicolson method, and the exact solutions of Example 1 in 2D for fixed $T = 0.5$. For $h = 0.1$, $k = 0.05$ and $T = 1$, the graphical representations of the space-time graph of the numerical solutions calculated by the DCNM and exact solutions for the above boundary value problem for $x \in [0,1]$ are presented in Figures 2 and 3, respectively, and it can be observed that both solutions are very similar.

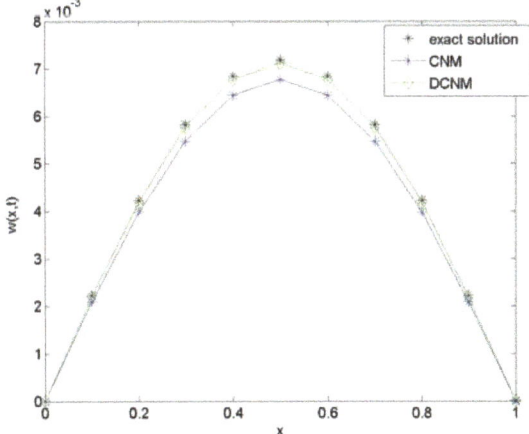

Figure 1. Comparison of the exact solution with solutions obtained by CNM and DCNM.

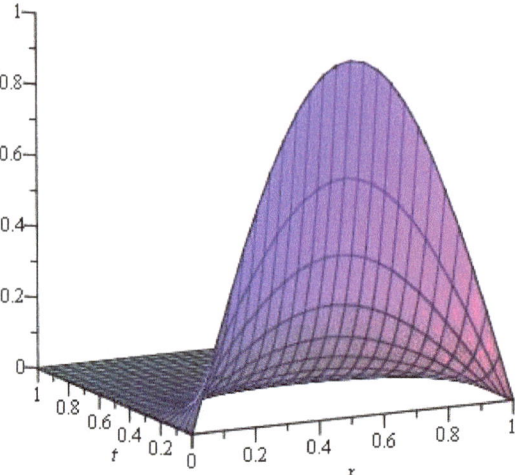

Figure 2. Space-time graph of DCNM solution for Example 1.

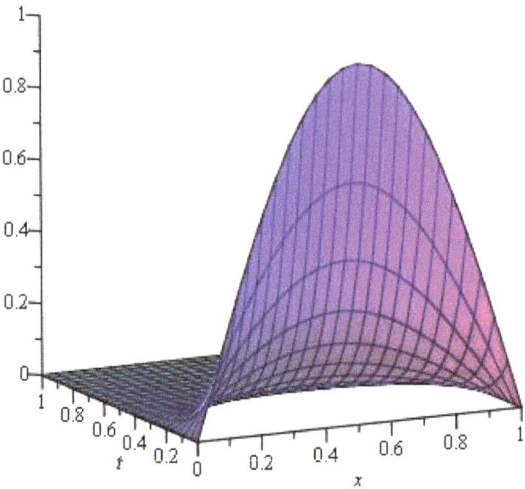

Figure 3. Space-time graph of exact solution for Example 1.

6. Conclusions

This paper contains the application of the discrete symmetry transformation for the boundary value problem of the diffusion equation. An invariantized finite difference method to find the solution of the heat equation using the composition of discrete symmetry group and projection group of the heat equation is developed. Tables 2–4 show that the proposed invariantized method DCNM improves the efficiency and accuracy of the existing Crank Nicolson method.

Similarly, with the help of discrete symmetry groups of partial differential equations, different invariantized finite difference schemes can be constructed to improve the efficiency and performance of the existing finite difference methods.

Author Contributions: All authors have equal contribution to this research and in preparation of the manuscript. All authors have read and agreed to the published version of the manuscript.

Funding: This research was funded by NUST, Pakistan.

Acknowledgments: The author would like to thanks the editor and anonymous referees for their suggestions and valuable comments that improved the manuscript.

Conflicts of Interest: The authors declare no conflict of interest.

Abbreviations

The following abbreviations are used in this manuscript:

FTCS   Forward in Time and Centered in Space
CNM   Crank Nicolson Method
DCNM   Discretized Crank Nicolson Method

References

1. Lie, S. *Theorie der Transjormations Gruppen*; Chelsea: New York, NY, USA, 1970.
2. Yang, H.; Shi, Y.; Yin, B.; Dong, H. Discrete Symmetries Analysis and Exact Solutions of the Inviscid Burgers Equation. *Discret. Nat. Soc.* **2012**, *56*, 1–15. [CrossRef]
3. Ibragimov, N.H. *Elementary Lie Group Analysis and Ordinary Differential Equations*; John Wiley & Sons: Chichester, UK, 1999.
4. Golubitsky, M.; Stewart, I.; Schaeffer, D.G. *Singularities and Groups in Bifurcation Theory*; Springer: New York, NY, USA, 1988.
5. Hydon, P.E. Discrete point symmetries of ordinary differential equations. *R. Soc. Lond. Proc. A* **1998**, *454*, 1961–1972. [CrossRef]
6. Hydon, P.E. How to construct the discrete symmetries of partial differential equations. *Eur. J. Appl. Math.* **2000**, *11*, 515–527. [CrossRef]
7. Hydon, P.E. *Symmetry Methods for Differential Equations*; Cambridge University Press: Cambridge, UK, 2007.
8. Laine, F.E.; Hydon, P.E. Classification of discrete symmetries of ordinary differential equations. *Stud. Appl. Math.* **2003**, *111*, 269–299. [CrossRef]
9. Garcıa, M.M.; Zhang, Y.; Gordon, T. Modelling Lane Keeping by a Hybrid Open-Closed-Loop Pulse Control Scheme. *IEEE Trans. Ind. Inform.* **2016**, *12*, 2256–2265. [CrossRef]
10. Garcıa, M.M.; Gordon, T. A New Model of Human Steering Using Far-Point Error Perception and Multiplicative Control. In Proceedings of the IEEE International Conference on Systems, Man, and Cybernetics (SMC2018), Miyazaki, Japan, 7–10 October 2018.
11. Kim, P. Invariantization of numerical schemes using moving frames. *Numer. Math. Springer* **2006**, *10*, 142–149. [CrossRef]
12. Olver, P.J. Geometric foundations of numerical algorithms and symmetry. *Appl. Algebra Eng. Commun. Comput.* **2001**, *11*, 417–436. [CrossRef]
13. Budd, C.; Dorodnitsyn, V.A. Symmetry-adapted moving mesh schemes for the nonlinear Schrodinger equation. *J. Phys. A Math. Gen.* **2001**, *34*, 10387–10400. [CrossRef]
14. Dorodnitsyn, V.A. Finite difference models entirely inheriting continuous symmetry of original differential equations. *Int. J. Mod. Phys. Ser. C* **1994**, *5*, 723–734. [CrossRef]
15. Dorodnitsyn, V.A.; Kozlov, R.; Winternitz, P. Lie group classification of second order difference equations. *J. Math. Phys.* **2000**, *41*, 480–504. [CrossRef]
16. Valiquette, F.; Winternitz, P. Discretization of partial differential equations preserving their physical symmetries. *J. Phys. A Math. Gen.* **2005**, *38*, 9765–9783. [CrossRef]
17. Budd, C.J.; Iserles, A. Geometric integration: Numerical solution of differential equations on manifolds. *Philos. Trans. R. Soc. Lond. A*, **1999**, *357*, 945–956. [CrossRef]
18. Verma, R.; Jiwari, R.; Koksal, M.E. Analytic and numerical solutions of nonlinear diffusion equations via symmetry reductions. *Adv. Differ. Equ.* **2014**, *10*, 142–149. [CrossRef]
19. Kim, P. Invariantization of the Crank-Nicolson method for Burgers' equation. *Phys. D Nonlinear Phenom.* **2008**, *237*, 243–254. [CrossRef]
20. Stavroulakis, I.P.; Tersian, S.A. *Partial Differential Equations: An Introduction with Mathematica and Maple*; World Scientific Publishing Company Ltd.: Hackensack, NJ, USA, 2004.

21. Bluman, G.W.; Kumei, S. *Symmetries and Differential Equations*; Springer: New York, NY, USA, 1989.
22. Burden, R.L.; Faires, J.D. *Numerical Analysis*, 9th ed.; Brooks/Cole Cengage Learning: Boston, MA, USA, 2010.
23. Kharab, A.; Guenther, R.B. *An Introduction to Numerical Methods, A Matlab Approach*, 3rd ed.; CRC Press Taylor and Francis Group: New York, NY, USA, 2011.
24. Marx, C.; Aziz, H. Lie Symmetry preservation by Finite difference Schemes for the Burgers Equation. *Symmetry* **2010**, *2*, 868–883.

© 2020 by the authors. Licensee MDPI, Basel, Switzerland. This article is an open access article distributed under the terms and conditions of the Creative Commons Attribution (CC BY) license (http://creativecommons.org/licenses/by/4.0/).

MDPI
St. Alban-Anlage 66
4052 Basel
Switzerland
Tel. +41 61 683 77 34
Fax +41 61 302 89 18
www.mdpi.com

Symmetry Editorial Office
E-mail: symmetry@mdpi.com
www.mdpi.com/journal/symmetry

www.ingramcontent.com/pod-product-compliance
Lightning Source LLC
LaVergne TN
LVHW070554100526
838202LV00012B/461